MAR   2012

ocean
drifters
a secret world
beneath the waves

RICHARD R KIRBY

ocean
drifters
a secret world
beneath the waves

RICHARD R KIRBY

FIREFLY BOOKS

A FIREFLY BOOK

Published by Firefly Books Ltd. 2011

Text and Images Copyright © 2010 Richard R. Kirby
Compilation © Studio Cactus Books

First printing

Publisher Cataloging-in-Publication Data (U.S.)
A CIP record of this book is available from Library of
Congress

Library and Archives Canada Cataloguing in Publication
A CIP record of this book is available from Library and
Archives Canada

Published in the United States by
Firefly Books (U.S.) Inc.
P.O. Box 1338, Ellicott Station
Buffalo, New York 14205

Published in Canada by
Firefly Books Ltd.
66 Leek Crescent
Richmond Hill, Ontario L4B 1H1

Printed in China

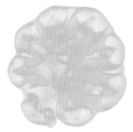

Contents

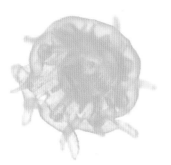

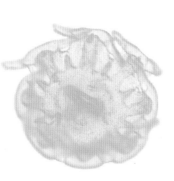
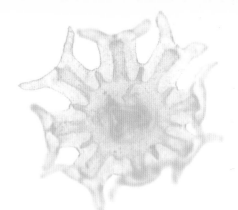

PHYTOPLANKTON BLOOM The swirls of surface foam stretching as far as the eye can see indicate a phytoplankton bloom has occurred in the water below.

Introduction

Through the eyepiece of a microscope, this book introduces the amazing variety of a hidden, rarely seen microcosm of life, the ocean drifters otherwise known as the plankton in the sea: microscopic phytoplankton – plant-like cells, many smaller than the diameter of a human hair – and the tiny animals that eat them, the zooplankton. These myriad creatures float freely in the sunlit surface of the sea, where they underpin the marine food chain, provide the world with oxygen, and – through their birth, growth, death and decay – play an essential role in the global carbon cycle. Without these remarkable creatures the oceans would be a barren wilderness, there would be no fish in the sea, we would have no reserves of oil or gas and the Earth's climate would be quite different. Although people are generally unaware of their presence beneath the sea's surface, the plankton create the characteristic smell of the sea that lures us to the seaside and they are even instrumental in cloud formation. At present, as their habitat alters rapidly due to rising sea temperatures as a consequence of global climate change, their distributions, abundance and seasonality are changing with ramifications for the whole marine food web and for the ecology of our entire planet.

Going with the flow

Plankton range in size from picoplankton smaller than 0.2 µm (0.0002 mm), the marine viruses – the virioplankton – and the free-living marine bacteria – the bacterioplankton – to organisms as large as the Nomura's jellyfish, *Nemopilema nomurai*, which can reach 2 m across and weigh more than 200 kg. In between these extremes of size is a vast diversity of life that includes phytoplankton – single-celled photosynthetic bacteria and microalgae – and the zooplankton – the animals that graze the phytoplankton – and

their planktonic predators in turn. The name plankton is derived from the Greek word *planktos*, which quite literally means a 'drifter', since the characteristic feature of all these creatures is that they travel at the whim of the ocean currents – and so by definition the plankton excludes animals such as fish that are strong enough to swim against a flow of water.

In contrast to the almost limitless horizontal distribution of the plankton, their vertical extent is constrained in the oceans mainly to the sunlit layer of the sea – the photic zone – which comprises the euphotic zone, where sufficient light penetrates to allow photosynthesis, and the darker disphotic zone below, which extends to where the intensity of light falls to 1% of its surface value.

Plankton and the marine food web

At the very base of the marine food chain, the phytoplankton are the ocean's 'primary producers'. Despite their small size the phytoplankton account for approximately 50% of all photosynthesis on the planet and, together with land plants, they are the principal producers of atmospheric oxygen, a by-product of photosynthesis. These figures are less surprising when you realise that the oceans and seas cover more than 70% of the globe's surface area. Although some species of phytoplankton live as chains of cells or large colonies, most exist as single-cells and they can occur in such vast numbers when they bloom that they are visible from space. It is at the level of the phytoplankton that the virioplankton play a particularly important role in the plankton food web. Viral infection of bacteria and microalgae can control the duration of phytoplankton blooms to influence the extent of primary production in the oceans.

The phytoplankton are grazed by the herbivorous zooplankton – primary consumers –

just like cows graze grass on land. These vegetarians range from simple, single-celled ciliate protozoa to complex multicellular animals such as filter-feeding doliolids, salps, shrimp-like euphausiids and copepods. In turn, the smaller herbivores are eaten by predatory zooplankton such as carnivorous copepods, crab larvae, siphonophores, jellyfish and arrow worms – known collectively as secondary consumers. This planktonic food web provides nutrition for fish larvae (such as the larvae of cod) and planktivorous fish such as the sandeel, herring, anchovy and sardine – eaten in turn by some of the largest predatory fish such as the tuna, by sharks, by fish-eating seabirds such as the puffin and the albatross, and by mammals such as dolphins, killer whales and seals.

Zooplankton are also the direct food source for one of the largest sharks in the sea, the impressive basking shark, *Cetorhinus maximus*, and for some of the world's largest mammals, the filter-feeding baleen whales such as the Southern Ocean right whale, *Eubalaena australis*, that feeds on the Antarctic krill, *Euphausia superba*, or the Arctic bowhead whale, *Balaena mysticetus*, that feeds on smaller calanoid copepods. They are also the natural food for many surface-feeding oceanic seabirds, such as fulmars, petrels and kittiwakes.

Plankton and the carbon cycle
Ultimately, it is this food chain, from plankton to top predators that reveals another important feature of the plankton: its role in the global carbon cycle. Through the food web, inorganic carbon, which originated as carbon dioxide dissolved in seawater, is transformed first by phytoplankton photosynthesis into organic carbon compounds in their tissues and then

HUMPBACK WHALE, HAWAII Whether feeding like a humpback whale on a mixture of small fish and plankton, or predominantly on plankton like the Arctic bowhead whale, these huge mammals rely on the smallest of creatures for their survival.

KITTIWAKES PLUNGE DIVING FOR FOOD IN SVALBARD, NORWAY As well as taking fish larvae, the kittiwake's diet also includes zooplankton. Even seabirds such as puffins that do not feed on plankton directly, rely on the plankton food chain.

BACTERIOPLANKTON Among the smallest of the plankton – each of these bacterial cells is only 0.01 mm - they can number up to 1 million cells per ml of seawater where they play a central role in the plankton food web by recycling nutrients.

transferred up the food chain. At this point, another critical component of the plankton food web involving the bacterioplankton comes into play: the microbial loop. The bacterioplankton feed on dissolved organic material, such as the liquid wastes of zooplankton and phytoplankton. The bacterioplankton then become the food of single-celled protozoa, which are in turn eaten by zooplankton. In this way, the microbial loop recycles nutrients within the planktonic food chain. The true importance of the microbial food web can be appreciated when it is realised that bacterioplankton productivity – the amount of bacterioplankton carbon biomass – can equal that of the phytoplankton.

When the plankton and larger animals die, their decaying bodies (together with the faecal pellets from living plankton) sink to the sea bed taking the carbon they contain with them. Often, the quantity of sinking planktonic material is so abundant in the water column that it is referred to as 'marine snow'. The bacterioplankton now play another important role in the marine food web. By decomposing the sinking organic matter as it descends to the sea bed, the bacterioplankton recycle some of the organic carbon, nitrate, phosphate and silicate to the water column. Eventually however, a proportion of the organic matter arising from primary production at the surface, reaches the sea floor making the deep ocean both rich in nutrients and a large reservoir of carbon.

Plankton Productivity

Although plankton can be found throughout the world's oceans, they are most abundant in nutrient-rich regions such as shallow seas, like the North Sea, on continental shelves and in upwelling regions where deep ocean water rises to the surface, such as the Antarctic polar fronts,

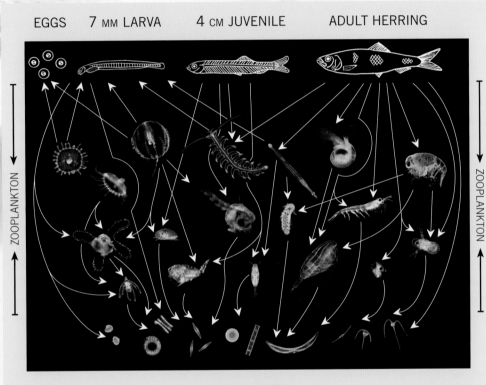

EGGS   7 MM LARVA   4 CM JUVENILE   ADULT HERRING

ZOOPLANKTON

ZOOPLANKTON

PHYTOPLANKTON

THE PLANKTON FOOD WEB WITH RESPECT TO HERRING

The intricacies of the North Sea plankton food web with respect to a planktivorous fish. Phytoplankton sit at the base of the food chain and are known as primary producers. The phytoplankton provide the food for the zooplankton primary consumers such as copepods like *Calanus helgolandicus* and euphausiids such as *Nyctiphanes couchii* along with the young planktonic larvae of many animals that live on the sea bed, such as those of decapods (crabs and shrimps) and bivalves (mussels and scallops). In turn the primary consumers are the food of predators, the secondary consumers such as hyperiid amphipods, decapod larvae, jellyfish and arrow worms. At the top of this food chain are larger animals such as the filter-feeding herring that depends on the plankton as a source of food throughout its entire life. Changes in the plankton such as subtle changes in the seasonality of individual species – the precise time of year that they occur – or in their abundance due to changes in sea temperature, have the potential to uncouple this food web with ramifications throughout the pelagic food chain. Plankton food web adapted from an original idea by Sir Alister Hardy FRS. Images not to scale.

the Benguela upwelling region to the west of South Africa, the coasts of Peru and Chile, and eastern New Zealand. Shallow seas and upwelling regions that are rich in nutrients can support high levels of phytoplankton primary production, which in turn promotes a diverse plankton food chain to make these regions some of the most productive fisheries in the world. Although upwelling water is nutrient rich, not all upwelling regions of the oceans are 100% productive, however. These less productive areas are called high-nutrient, low-chlorophyll (HNLC) regions and it is thought that primary production is limited by the availability of phosphate (in the North Atlantic) and iron (in the equatorial and Southern Pacific Oceans) in the water.

Life in the plankton

Not all planktonic species spend their whole life in the plankton. In fact, scientists separate the zooplankton into two categories: those animals that complete their whole life cycle, from egg to adult, in the plankton – called the holozooplankton – and those that spend just a part of their life cycle in the plankton – known as the merozooplankton. The merozooplankton covers the larval stages of animals that live on the sea bed, such as crabs, worms, molluscs and sea urchins. The benefit of a planktonic larval stage is considered to be twofold. Firstly, the productive sea surface is a good place to feed, and secondly, the ocean currents provide a mechanism that enables species that are

sedentary on the sea floor to disperse – often over long distances – to new locations.

For most of these organisms, whether phytoplankton, holozooplankton or merozooplankton, their life in the plankton is usually short. Phytoplankton live for only a few days or weeks, either dying when they are eaten, after they become infected by viruses or when nutrients are insufficient to support their growth. While the zooplankton may live longer, up to well over a year in the case of many euphausiids, hyperiids and large cold-water copepods, their lives are often ended abruptly by predation. For all holozooplankton the goal is to survive long enough to reproduce, while for merozooplankton it is to survive the planktonic phase and find

JUVENILE CRAB A young edible crab, *Cancer pagurus,* seeks shelter among seaweed and limpets on the seashore. This crab started life in the plankton where, as a zoea larva, it fed first on phytoplankton and then as a predator on other zooplankton before settling to the sea bed. The planktonic larval stage allows the crab to disperse to new locations.

CRAB POTS Sitting on the harbourside, these pots that are used to catch crabs and lobsters on the sea bed would remain empty if the animals they catch had not first completed their larval stage in the plankton.

somewhere to settle and become an adult. To aid their survival, zooplankton have therefore evolved several ways to sense their environment. Most zooplankton have some form of photoreceptor, varying from simple ones to detect light and motion (the latter may help them to avoid predation) to complex eyes, which, in the case of some predatory species, are used to locate their prey. Several animals also possess sensory bristles and antennae that are both mechano-sensory – able to detect vibrations – and chemo-sensory – able to detect smells. The antennae of copepods are particularly important in helping these animals find a suitable mate from among all the other zooplankton species. Copepods do this first by following a scent trail in the water and then by using their antennae to explore the surface of the potential partner to detect species-specific scents called pheromones.

Migration

Although the plankton are at the mercy of ocean currents and cannot swim against a horizontal flow of water, they can travel great distances vertically in relation to their size by either swimming or adjusting their buoyancy. This vertical migration can occur on a daily basis, or seasonally. For example, motile phytoplankton, such as dinoflagellates, may descend a few metres at sunset to more nutrient rich waters, rising back to the surface at sunrise. The greatest vertical migrations, however, are seen among the zooplankton. Many zooplankton spend the daylight hours in the relative safety of the darkness of the disphotic zone where they are less visible to predators, only rising to the surface at night to feed on the phytoplankton. The 4 mm-long copepod *Pleuromamma robusta* can descend to over 400 m daily by swimming and

adjusting its buoyancy; this represents a journey of over 100,000 times its own body length. The energetic costs associated with this behaviour indicate how important it is for their survival that they spend the daylight hours in darkness.

On a seasonal basis, many copepods diapause over winter in dense aggregations and at great depths, prior to rising to the surface in spring where subsequently they moult into adults and reproduce to complete their life cycle. Diapause is a physiological state of dormancy in which every cell ceases growth thus enabling copepods to survive during predictable, unfavourable conditions such as the winter. The North Pacific copepod *Neocalanus cristatus* sinks to depths of up to 2 km in winter, equivalent to a human climbing from Mt Everest base camp to the summit over 140 times. Diapause has very specific triggering and releasing conditions. It is thought that the spring awakening of copepods is triggered by a combination of rising sea temperatures and increasing day-length, the latter detected through subtle changes in the small amount of light that reaches deep waters. Whether at the surface or deeper, the growth and activity periods of most plankton are set in motion by this seasonal change in temperature and/or day-length, and they terminate typically either when surface nutrients are depleted or when day-length and temperature reduce in the autumn.

Planktonic life between the sea and air

Pleuston is the name given to the group of oceanic animals that live at the sea–air interface, half in and half out of the water. Some of these animals such as the Portuguese man o' war, *Physalia physalis*, the by-the-wind sailor, *Velella velella*, and the violet snail, *Janthina janthina*, are planktonic, travelling at the mercy of the

THE GAS-FILLED BLADDER OF A PORTUGUESE MAN 'O' WAR Blown onto a beach by the wind, this is often where you may find this jellyfish stranded. Even when lying on the beach its trailing blue tentacles may still deliver a powerful sting.

ocean currents and the wind. To help them stay afloat at the surface they all have some form of buoyancy device. The *P. physalis* has a gas-filled, balloon-like bladder that sits above the sea surface, *V. velella* has a flat disc comprising a series of concentric air-filled tubes above which a stiff, sail-like vane projects upwards into the air, and *J. janthina* secretes a raft of air-filled mucous bubbles beneath which the snail hangs. The gas-filled bladder of *P. physalis* and the sail of *V. velella* are orientated obliquely with respect to the animal's longitudinal axis and so cause them to drift at 45 degrees to the prevailing wind. There are two distinct forms of *P. physalis* and *V. velella* that are distinguished by the orientation of the bladder or the sail, respectively; in both species there are variants where the bladder or the sail are orientated either to the left or to the right of the long axis.

By having two orientations it helps ensure these animals disperse widely and evenly across the oceans where they live.

The Portuguese man o' war and the by-the-wind sailor are both Cnidaria – jellyfish – and so they possess nematocysts – stinging cells – that they use to catch their food. The larger *P. physalis* feeds mainly on young fish whereas the smaller *V. velella* feeds on fish eggs, fish larvae and zooplankton. Both *P. physalis* and *V. velella* also have a truly planktonic larval stage, in contrast to *J. janthina* that gives birth to live young. In *P. physalis* reproduction occurs by the release of either sperm or eggs into the water from specialised cells called gonozooids which, after fusion, give rise to the planktonic larva directly. In *V. velella* the gonozooids produce small free-living male or female medusae by asexual budding. The sexual medusae of *V. velella* then liberate eggs or sperm to give rise to planula larvae that live at depths of up to 2,000 m where ocean currents that are quite different to those at the surface disperse them to new locations. As the *V. velella* larvae grow they accumulate oil droplets that make them rise eventually to the surface where they mature into the adult jellyfish. Because the pleuston are blown by the wind, they can appear on beaches after persistent, strong onshore gales; the by-the-wind sailor may be washed up on the strandline in enormous numbers.

The snail *J. janthina* is immune to the nematocysts of *V. velella* and *P. physalis* upon which it feeds. As a form of camouflage, *P. physalis*, *V. velella* and *J. janthina* are all coloured deep blue. In the case of *J. janthina*, however, the blue coloration is confined to the surface of the shell that faces skyward. The side of the shell that faces the ocean's depths (this is the top of the animal as *J. janthina* hangs upside down) is almost white. This counter-shading counteracts the animal's shadow providing camouflage from predators such as fish and turtles that attack it from below.

Harmful algal blooms

While most plankton are beneficial and harmless, a few phytoplankton species can harm food webs when they bloom and reach very high numbers in the water. Referred to as harmful algal blooms (HABs) when they occur, many simply suffocate marine life by causing oxygen depletion in shallow seas when they die and decay. Some species, however, such as diatom species that have sharp spines, may suffocate fish directly by lodging in their gills and causing acute inflammation. Many other harmful algae produce potent toxins that can accumulate through the food chain causing massive fish kills, and bird and mammal mortalities. Humans can be affected by this toxin build-up too if they eat shellfish containing high numbers of toxic algae, which can lead to respiratory, neurological and gastrointestinal disorders, and even death in extremely rare circumstances. Because the aquaculture industry (fish and shellfish farming) is especially vulnerable to HABs, many coastal waters are closely monitored to provide an early warning system for the presence of harmful algae and the potential development of a HAB. In this way, the likelihood of toxic impacts on humans is minimised.

Lighting up the plankton

A wide range of planktonic creatures such as dinoflagellate phytoplankton, some species of copepods, euphausiids, ostracods, jellyfish and siphonophores can produce light by a chemiluminescent reaction in a phenomenon known as bioluminescence. In most cases this involves the oxidation of a protein called luciferin in a reaction catalysed by the enzyme luciferase. Unlike the firefly on land that uses bioluminescence to attract a mate with flashes of light, it is not truly known why most planktonic creatures bioluminesce. In many cases they produce short flashes of light of less than a second in duration. These flashes are usually in response to an external stimulus – typically when they are disturbed or attacked – and so they are thought to be a defence mechanism. Bioluminescent dinoflagellate blooms are especially spectacular and can make the whole surface of the sea appear to sparkle with light at night; bioluminescent dinoflagellates are responsible for the luminescence in breaking waves or the wake of a ship, for example. The light emitted by dinoflagellates such as *Noctiluca miliaris* (*noct* meaning night and *luc* meaning light) is thought to both startle the creatures that eat them (copepods and euphausiids) and to act as a 'burglar alarm' that attracts predators of the copepods and euphausiids towards the dinoflagellate bloom.

Defence mechanism

The light produced by copepods such as *Metridia lucens* or by jellyfish such as *Pelagia noctiluca* may also serve to deter predators. Likewise, the ostracod *Cypridina hilgendorfii* may use bioluminescence for defence, but in this case more akin to the way that an octopus might use its ink to escape a predator. When *C. hilgendorfii* is disturbed, it releases a puff of luciferin and luciferase into the sea. When these two molecules mix in the water they produce a confusing cloud of blue light enabling the ostracod to swim away into the darkness.

Light may not just be produced as a defence mechanism when an organism in disturbed, however. For example, euphausiids have highly developed compound eyes and, like many of these shrimp-like plankton, *Meganyctiphanes norvegica* emits light from specialised organs called photophores (one on each eye stalk, four on the thorax and four on the abdomen). These bright pinpoints of blue light may enable *M. norvegica* to communicate with each other. While it is possible to speculate why dinoflagellates and some zooplankton are bioluminescent, it is not known why some of the bacterioplankton emit light. Unlike dinoflagellates and zooplankton that bioluminesce only when disturbed, some bacteria such as *Vibrio harveyi* bioluminesce when they reach a high concentration in the water. This bioluminescence, which does not appear to be in response to external stimuli, can give the sea a night-time milky glow of pale blue light that can stretch from horizon to horizon and be visible from space.

Perhaps the most famous bioluminescent member of the plankton, however, is the jellyfish *Aequorea victoria*. This animal uses the bioluminescent protein aequorin, which emits blue light when it binds with calcium. Although aequorin emits blue light the jellyfish actually emits green light because the blue light is immediately absorbed by another protein called green fluorescent protein (GFP). It is GFP that has made *A. victoria* famous since its discovery, isolation and the subsequent cloning of the GFP gene led to a Nobel Prize in chemistry in 2008. GFP has enabled large advances in science and biomedical research through its use as a fluorescent protein label.

Plankton and climate

As well as supporting the marine food chain, the plankton have an important influence upon climate through their effect on the atmospheric

VISIBLE FROM SPACE This satellite image was taken approximately 784 km above the surface of the Earth by the MEdium Resolution Imaging Spectrometer (MERIS) sensor on the European Space Agency (ESA) Envisat satellite, launched in 2002. The white swirls in the sea to the west and northwest of Ireland are a characteristic feature of a coccolithophore phytoplankton bloom.

concentration of the greenhouse gas carbon dioxide. The plankton achieve this in two ways and on two timescales. On a long-term timescale – over millions of years – the dead remains of plankton that sink to the sea bed lead to the sequestration of carbon dioxide as organic carbon in sedimentary rocks (some of which becomes oil and gas when it is heated and compressed) and as inorganic carbon in carbonate rocks, such as the instantly recognisable 80 m-high White Cliffs of Dover in the UK. It is incredible to realise that the accumulated dead remains of such minute creatures created these deposits over hundreds

of millions of years. On shorter-term timescales – over hundreds to a few thousand years – the plankton influence the amount of carbon that is sequestered in a carbon 'reservoir' in the waters of the deep ocean.

Carbon capture

The formation of inorganic carbonates by plankton actually links life in the sea with rock weathering and carbon dioxide uptake on land. On land, microbial decomposition of plant material in the soil creates a build up of organic acids and carbon dioxide. Rainwater is also slightly acidic due to dissolved carbon dioxide,

and, together with the organic acids, they bring about the weathering of rock minerals such as calcium silicate or magnesium silicate. In the case of calcium silicate, two molecules of carbon dioxide combine with the rock mineral to produce calcium, bicarbonate and silicic acid. After these weathering products are transported by river to the sea they are precipitated biogenically by the plankton. Calcium carbonate is precipitated by microalgae called coccolithophores in the form of their external calcium carbonate plates known as coccoliths, and by amoeboid foraminifera as calcium carbonate in their shells. The silicic acid is precipitated by diatoms to form their external silicaceous cell walls – the frustule – and by amoeboid radiolarians as their silica skeleton. This entire process results in the net removal of

WHITE CLIFFS The author standing at the base of one of the Seven Sisters. These massive white cliffs of calcite on the south coast of England are made almost entirely of the remains of single-celled coccolithophores deposited approximately 145 to 65 million years ago. The deposits were created during the Cretaceous period of Earth's history and later exposed by a change in sea level. Deposits such as these are part of the Earth's long-term carbon cycle. These calcium carbonate cliffs represent the sequestration of carbon dioxide removed from the atmosphere during the process of rock weathering and its subsequent biogenic precipitation by the plankton. Naturally, the carbon sequestered in these rocks is returned slowly to the atmosphere by weathering on land. The process of cement manufacture from calcium carbonate rocks liberates carbon dioxide into the atmosphere and represents a rapid acceleration of the natural process.

carbon dioxide from the atmosphere during the weathering process and its sequestration by the plankton in the geological or rock carbon reservoir. In effect, carbon dioxide from the atmosphere is converted into the bodies of plankton, which sink to the sea floor when they die, and, over long geological timescales, their bodies become rocks. It is a long-term net removal of carbon dioxide, since for every two molecules of carbon dioxide used in the weathering of calcium silicate on land, only one of them is incorporated into calcium carbonate by the plankton, the other is given off to the seawater during the biogenic process. In this way, through the liberation of carbon dioxide, changes in the rate of biogenic calcium carbonate production in the oceans can influence the short-term carbon cycle described below.

Under natural conditions on long-term time scales, the sequestered inorganic carbon is returned to the atmosphere by degassing of carbon dioxide at volcanoes and at mid-ocean ridges, or when limestone is subducted and melted or exposed and weathered. Similarly, buried organic carbon from the soft tissues of dead plankton that have sunk to the sea floor, is liberated by rock weathering or thermal decomposition. Our burning of fossil fuels and manufacture of cement reflect a rapid acceleration of this natural regeneration process and it is removing the geological carbon reservoir 100 times faster than would occur naturally, liberating the greenhouse gas carbon dioxide into the atmosphere in the process.

Plankton and the short-term carbon cycle

The plankton also play a central role in the global carbon cycle on more short-term timescales. Because the oceans are in contact with the atmosphere, carbon dioxide enters the sea surface from the air. Compared with the atmosphere that contains about 730 Gt of carbon (1 giga tonne (Gt) = 1 billion metric tonnes), the oceans contain about 38,000 Gt of carbon, of which 97% resides in the deep ocean below 100 m. When carbon dioxide dissolves in seawater it forms a weak acid – carbonic acid – that breaks down quickly to produce bicarbonate and hydrogen ions. Consequently, only 0.5% of the inorganic carbon in seawater occurs as carbon dioxide gas, which favours the further uptake of carbon dioxide from the atmosphere.

Carbon dioxide dissolves much more easily in cold water and so uptake is greatest at high latitudes near the poles. Because cold polar water is also denser, it sinks and transfers the carbon it contains to the deep ocean where it then circulates – this process is called the physical carbon pump or, the solubility pump. Eventually, many hundreds of years later, this carbon-rich deep water, which is now also nutrient rich, returns to the surface at upwelling regions. Here, high levels of primary production at the surface consume carbon dioxide and the sinking dead organic matter (marine snow) returns, or pumps, carbon back to the deep ocean. This process, which is referred to as the 'biological carbon pump', consequently works against the physical upwelling of carbon to continually return carbon to the deep ocean reservoir. Together, the physical pump and the biological pump control the carbon budget in the ocean and hence influence the atmospheric concentration of carbon dioxide, and therefore climate, over short-term timescales (hundreds of years). It has been estimated that carbon dioxide levels in the atmosphere would almost double from present levels if primary production and the biological carbon pump in the oceans ceased; this would be due to the dissolved

FOAMING PLANKTON Foam that is often blown onto beaches in spring and autumn is formed from the decaying remains of a *Phaeocystis* phytoplankton bloom. *Phaeocystis* spp., live as large colonies within a mucilaginous (jelly-like) matrix made of sugars and, when the bloom dies, their dead remains can be whipped into a foam by wave action.

carbon in the deep oceans equilibrating with the atmosphere. Conversely, it has been estimated that if all the HNLC regions of the oceans were 100% productive the atmospheric carbon dioxide concentration would more than halve from present levels.

Natural climate change

Over the last 800,000 years, prior to the Industrial Revolution at the beginning of the 19th century, ice-core evidence suggests that atmospheric carbon dioxide levels fluctuated between 172 parts per million (ppm) and 300 ppm, and that these fluctuations coincided with the glacial–interglacial periods. Changes in planktonic primary production in the oceans are thought to be one of the main influences on atmospheric carbon dioxide levels during this time, via a series of positive and negative feedback mechanisms. One of the feedback mechanisms that might have led to a reduction in atmospheric carbon dioxide, and so a cooling climate during glacial periods, is believed to involve the fertilisation of the oceans with phosphate and other micronutrients, especially iron. As the glacial period begins, a steep temperature gradient develops between the freezing poles and the warmer equator that strengthens mid-latitude winds that transport dust from deserts to the oceans. It is suggested that these aeolian inputs – wind-borne sediments – promote increased primary

POLAR BEARS At the top of the Arctic marine food chain, polar bears feed mainly upon ringed and bearded seals whose diets include fish, and marine invertebrates such as shrimp and squid.

production and a more powerful biological carbon pump leading to a drawdown of atmospheric carbon dioxide into the sea. In fact, scientists have suggested recently, that one way to help ameliorate climate change due to rising carbon dioxide levels in the atmosphere might be to promote the biological carbon pump in the oceans. Their idea is to fertilise the sea surface artificially with iron in order to encourage the growth of microalgae and enhance the drawdown of carbon dioxide from the atmosphere.

Current climate change and plankton

While the plankton are intimately connected with the Earth's climate through their pivotal role in the global carbon cycle, they are also influenced by climate. As humans, we often forget that most organisms on Earth, such as the plankton, are cold blooded – ectothermic – and so their metabolism is influenced directly by the temperature of their environment. Living in the surface of the sea, the plankton are particularly sensitive to changes in sea surface temperature. The level of carbon dioxide in the atmosphere has risen over the last 200 years from approximately 280 ppm to approximately 380 ppm in 2008, due largely to the burning of fossil fuels and cement manufacture. This rate of change is 100 times greater than the most recent natural change in atmospheric carbon dioxide at the end of the last ice age, and it has given carbon dioxide levels that are now 27% higher than at any point in the last 800,000 years. As a consequence of this anthropogenic (manmade) increase in atmospheric carbon dioxide, climate models predict that the global average temperature will rise by 0.4°C over the next 20 years, and between 2 and 4°C over the next century.

Global sea surface temperatures are already

1°C higher than 150 years ago and coincident changes are being observed in the abundance, distributions and seasonal timing of the plankton in both the Atlantic and Pacific oceans. We already know that normal year-to-year variations in sea surface temperature affect the plankton. In the Northeast Pacific, for example, both the timing of the peak in plankton biomass and its duration are influenced by sea surface temperature, occurring a month earlier in warm years and for a shorter period. This change in timing has implications for predators such as fish and seabirds that may time their migrations and breeding to match with the plankton. Likewise, across the whole northern North Atlantic, the abundance of the Arctic copepod *Calanus hyperboreus* is affected by year-to-year changes in the North Atlantic Oscillation (NAO), which is the dominant mode of winter climate variability in the region. The NAO affects winds and storm tracks across the North Atlantic, thereby altering air temperature, sea surface temperature, and precipitation. Changes in the NAO create a seesaw in air temperatures between the Northeast and Northwest Atlantic regions, which influences sea temperatures. This affects the year-to-year trans-Atlantic abundance of *C. hyperboreus* that, as the largest and most abundant copepod in the northern North Atlantic, is a critical food source for fish, birds and whales. When this copepod is abundant in the Northwest Atlantic it is less abundant in the Northeast Atlantic, and vice versa.

Changing Antarctic and Arctic conditions

One of the World's fastest warming seas due to global climate change is the western Antarctic Peninsula in the Southern Ocean. Here, the potential habitat for the planktonic krill lies between the polar front and the Antarctic ice shelf. Juvenile krill overwinter by feeding on sea ice algae and so their food supply is determined by the geographical extent of the sea ice. As winter sea ice extent has declined as the sea has warmed, so has krill recruitment – the number of juveniles reaching adulthood – declined. This has led to a two-fold decrease in the abundance of krill, which are key prey species for penguins, albatross and whales. Ironically, this decline in krill is occurring at the very time that whale numbers are recovering due to the moratorium on hunting.

In 1998, climate warming created extensive ice-free water to the north of Canada causing changes in Arctic hydrography and circulation that led to Pacific surface water flowing into the northwest North Atlantic. Coincident with this event, large numbers of the Pacific diatom *Neodenticula seminae* appeared in Atlantic plankton samples for the first time in over 800,000 years; previously the most recent records of this species in the North Atlantic were obtained from deep-sea sediment cores dated to the Pleistocene period, approximately 1.2 to 0.8 million years ago. It has been predicted that these trans-Arctic migrations of plankton species from the Pacific into the Atlantic may increase as Arctic ice continues to melt due to climate change.

Perhaps the most striking changes in distributions of the plankton in response to changes in sea temperature have been observed in samples collected in the Northeastern Atlantic by the Continuous Plankton Recorder (CPR) survey, which has been monitoring the North Atlantic plankton on a monthly basis since the 1930s. Analysis of CPR data has revealed that during the last 50 years there has been a northward movement of warm water copepod species by 10° latitude and a similar retreat of

cold water species towards the Arctic; this represents a mean poleward movement of plankton of between 200–250 km per decade.

Seasonal changes

In addition to geographical changes, the phenology, or the seasonal timing of occurrence of the plankton, is also changing as the seas warm. In the North Sea, many merozooplankton larvae now occur earlier in the year than before; the timing of the peak abundance of echinoderm larvae has advanced by more than 47 days over the last 50 years. Both biogeographical and phenological changes can have profound effects on the plankton food web. For example, the survival of larval cod in the North Sea is influenced by the timing, the quality and the quantity of their plankton food, especially the presence of their favoured cold-water copepod, *Calanus finmarchicus*. As the North Sea has warmed, and plankton distributions have moved northward, *C. finmarchicus* has declined by 70% in its abundance in the North Sea where it has been replaced by its southern, warm-water sister species, *C. helgolandicus*. Unfortunately, for larval cod, *C. helgolandicus* occurs in the plankton at a different time of year creating a trophic mismatch between the fish larvae and their favoured prey – the food for larval cod is not there at the right time. This change in the plankton affects cod recruitment and, consequently, the numbers of adult cod in the sea available to be caught by fishermen. The change in the North Sea plankton is not only affecting cod, however. The copepod, *C. helgolandicus* is of a poorer nutritional quality than *C. finmarchicus* and this has been linked with a decline in seabird populations such as the guillemot, *Uria aalge*. Changes in the nutritional quality of the copepod species being consumed

by the fish prey of guillemots is thought to be responsible for the decline in the bird's breeding success as the fish are both fewer and smaller.

Synergistic effects of fishing and climate on the plankton food web

The interactions within the plankton food web – between the different trophic levels represented by primary producers, herbivores, carnivores and detritivores – and between the plankton and the benthos (organisms living on the sea floor) impart stability to the ecosystem. Perturbations such as an environmental change, invasion by an alien species or the removal of a keystone species such as a top predator by, for example, overfishing, can alter the strength of the interactions among species and change the nature of the ecosystem, often abruptly.

Commercial fishing has depleted fish stocks worldwide and is held responsible for a shift in marine ecosystems from large predatory fish to short-lived small planktivorous fish, benthic invertebrates and jellyfish. In most cases, the target fish of commercial fisheries are top predators (top predators such as seals and sharks have long since declined due to hunting). One example is the effect of the overfishing of cod and other benthic fish such as haddock in the North Atlantic and adjacent seas. A large component of the diet of adult cod is decapods (crabs and shrimps) and small pelagic fish such as sprats. On the Scotian Shelf, the decline of cod during the 1980s and early 1990s coincided with an increase in northern shrimp, *Pandalus borealis*, and the snow crab, *Chionoecetes opilio*, and an increase in small pelagic fish. Coincident changes were also observed in the plankton. Large-bodied zooplankton that are eaten by small pelagic fish and by decapod larvae declined in abundance and phytoplankton increased in abundance, presumably due to decreased grazing pressure as result of a smaller herbivorous zooplankton population. Recently, similar effects have been observed in the Baltic Sea following the decline of cod. In the Baltic Sea, however, another manifestation of the decline of cod is thought to be an increase in the frequency of harmful algal blooms.

In the North Sea, overfishing of cod during the 1980s also coincided with a sustained increase in sea temperature due to a climate-driven warming; the North Sea is now 1°C warmer than in the late 1960s. This change in sea temperature appears to have acted synergistically with overfishing to bring about an abrupt change in the whole ecosystem. As described earlier, cod recruitment in the North Sea has been influenced by climate-induced changes in the holozooplankton. Studies of plankton data collected by the CPR survey reveal that the numbers of decapod and echinoderm larvae have increased from the mid 1980s while numbers of bivalve larvae in the North Sea plankton have declined to their lowest levels since records began. The larval abundance of decapods is greater in warm years than in colder years and so overfishing and changes in temperature may have both favoured decapod

abundance. Crabs are also important predators of bivalves affecting their recruitment, which may explain the decline in their larvae in the plankton. Phytoplankton has increased in the North Sea, similarly to the Scotian Shelf and the Baltic Sea. The most abundant echinoderm larvae in CPR samples of the North Sea is the heart urchin, *Echinocardium cordatum*, which as an adult, feeds on dead and decaying phytoplankton that sink to the sea floor. Warmer temperatures also favour the larval abundance of *E. cordatum*, and so both increased temperature and food supply may have benefited this species. The most recent change to be observed in the North Sea plankton is a proliferation of jellyfish. Since jellyfish can exert top-down and bottom-up control of fish larval survival, this may now signal the climax of the fishing and temperature-induced changes in the North Sea, and the establishment of a new ecological regime. Clearly, understanding the plankton food web is going to be essential to realising how marine ecosystems may respond to the synergistic effects of climate change and fishing.

Ocean acidification
In addition to climate change, the most recent concern to arise from increasing atmospheric concentrations of carbon dioxide is the effect on ocean acidity, or pH. As carbon dioxide increases in the atmosphere more carbon dioxide enters the sea surface. The additional hydrogen ions created by the dissociation of carbonic acid affect the pH and, consequently, the ocean surface has become more acidic, decreasing by about 0.1 pH units since the beginning of the Industrial Revolution. By the year 2100 it is predicted that the sea surface pH will decrease by a further 0.5 pH units, representing a 3-fold increase in the

SEABIRDS SURROUNDING A FISHING TRAWLER A flock of gannets feeds opportunistically on discards from a trawler. The plankton food web supports both pelagic and demersal fisheries for species such as herring, cod and tuna, halibut and plaice.

concentration of hydrogen ions. These changes in pH are so recent that it is still unclear how they will affect planktonic organisms, many of which produce the calcium carbonate mineral, calcite, such as coccolithophores and echinoderm larvae. Animals most susceptible to changes in pH, however, are likely to be those that produce an alternative form of calcium carbonate called aragonite, such as pteropod molluscs that have fragile aragonite shells. Pteropods are a major component of the Antarctic planktonic food web and are sometimes more important than krill as a food source. Aragonite dissolves readily and any decline in pteropod abundance as a result may influence the whole Antarctic food chain from fish through to penguins and seals.

Plankton and cloud formation

Although you may never have seen the plankton, if you have been to the coast, sailed a boat or taken an ocean cruise, you will have experienced the salty smell of the sea, sometimes called the 'sea air'. This characteristic smell is a result of volatile compounds produced by single-celled phytoplankton such as dinoflagellates and prymnesiophytes. One of the principal compounds is the chemical dimethyl sulphoniopropionate (DMSP), which is thought to protect them against salinity changes, to act as an antifreeze and to allow them to tolerate high levels of solar radiation. When the microalgae die, the DMSP is released into the seawater where it is degraded by bacteria and other phytoplankton into insoluble dimethyl sulphide (DMS). When DMS diffuses into the atmosphere it creates the smell that we associate with the sea. In the upper atmosphere, DMS is oxidised to particulate sulphate aerosols that attract water molecules and act as cloud condensation nuclei

to create clouds.

The transfer of DMS from the sea to the air above increases with increasing sea surface temperature. Because clouds are white they alter the Earth's albedo – its reflectivity – and consequently the amount of cloud cover affects the Earth's radiation balance and temperature. Since marine DMS production represents about 50% of the global biogenic source of sulphur, any increase in atmospheric DMS could provide a negative feedback loop at a time of global climate warming. Understanding the plankton and the biogeochemical cycles in the oceans is likely to prove central to understanding earth system science, especially at a time of climate change.

Plankton problems and solutions

In addition to the problems associated with harmful algal blooms, the larval forms that live in the plankton can also cause inconvenience if they choose to settle on underwater structures rather than the sea bed, such as on the hull of a boat or in the cold seawater intake of a power station. Known as biofouling, mariners are particularly aware of the problems caused by barnacles, tunicates and macroalgae (seaweeds) due to the effects of increased drag on the ship's hull. In fact, it is the fouling of a ship's hull that made sailors fear the punishment of keelhauling during the 18th and 19th centuries. Today, however, it is the costs to the shipping industry that are of more concern to sailors due to the fact that a fouled ship can burn up to 40% more fuel to achieve the same speed. In an attempt to stop the unwanted settlement of marine larvae, a large antifouling industry has developed. The oldest known methods to prevent the settlement of marine larvae are the use of copper or lead cladding by the Phoenicians and the Greeks and Romans,

COSTLY BARNACLES While these *Chthamalus montagui* are growing on the shore, barnacle larvae will also settle on other underwater surfaces such as ships' hulls. Called biofouling, this can cause a ship to burn up to 40% more fuel than when the hull is clean.

respectively. More recently, in the late 20th century toxic paints were developed that included the biocide tributyltin (TBT). Unfortunately, while TBT is a very potent biocide and antifouling coating, it was also found to have very harmful and persistent effects in the marine environment for non-target species. These effects were first noticed when the commercial production of oysters in Arcachon, France, almost collapsed. Soon a number of other deleterious effects were observed in other marine species. In particular, it was noticed that females of the intertidal snail *Nucella lapillus* developed male sex organs and became sterile in a process known as 'imposex'. Consequently, due to the ecological problems associated with

TBT its use is now banned in the marine environment worldwide. Today, antifouling technology is returning to more ingenious ways of using copper or by developing self-polishing coatings for ship hulls.

Plankton alien invaders

Although plankton can travel great distances by ocean currents they can also arrive at new places when we transport them in a less 'natural' way. The movement of marine organisms around the world by humans probably began as soon as we took to the seas. For example, fouling organisms like barnacles and seaweeds that hitched a ride on the hull of a ship will have given rise to planktonic larvae capable of establishing a new population if the habitat was favourable at the 'port of call'. The global spread of the notorious Japanese seaweed *Sargassum muticum*, which is very troublesome in harbours due to its rapid growth that clogs waterways, is thought to have occurred in this way.

As well as travelling on the outside of ships, planktonic organisms may also spread to new habitats by transport in a ship's ballast water or when new waterways are opened, such as the Panama or Suez canals that connect otherwise isolated habitats. While transport in ship ballast water is less effective for the dispersal of planktonic larvae – after a long journey most merozooplankton are likely to have lost their competence (ability to settle) – many phytoplankton and holozooplanktonic organisms may survive long distance travel. It is estimated, for example, that as a result of this transport of organisms around the world, there are now more than 400 non-indigenous species living on the shores and in the estuaries of the United States.

Often when alien invaders colonise a new habitat they can cause large ecological problems if they have no natural predators. It is thought that the ctenophore *Mnemiopsis leidyi* was introduced accidentally via ship's ballast water into the Black Sea in the 1980s. A native species of the eastern Atlantic, *M. leidyi* is a predator of zooplankton, fish eggs and fish larvae. Just prior to its arrival, overfishing in the Black Sea had removed most of the planktivorous fish such as sprats and anchovies that might have controlled its abundance. Consequently, without any natural predators, *M. leidyi* increased to large numbers in the plankton and consumed much of the zooplankton that fish larvae also rely upon for their food. Within a decade of the first noted appearance of *M. leidyi* in the Black Sea, the remaining fisheries had collapsed. Since its first introduction, *M. leidyi* has now spread naturally from the Black Sea to the Caspian and Mediterranean Seas, and most recently into the North Sea and the Baltic where there are new concerns for the effect it may have on these fisheries. Ironically, in the meantime, it is hoped that another recent introduction to the Black Sea may now act as a biological control of *M. leidyi* and help restore the ecosystem. In 1997, the ctenophore *Beroe ovata*, a preferential predator of other ctenophores, appeared in the Black Sea for the first time and already the abundance of *M. leidyi* is declining.

Because ports are often situated at the

STARFISH AND MUSSELS This young *Asterias rubens* starfish will feed upon the mussels it is sitting above by turning its stomach inside out to engulf its prey. While both these animals live on the sea bed as adults, they began their lives as larvae in the plankton.

ALIEN INVADERS The transport of planktonic organisms around the world in ship ballast water has caused the global spread of several species. Today, ships replace their ballast waters at sea to prevent the transport of coastal species to new locations.

mouths of estuaries, these areas are especially vulnerable to invasion by non-indigenous species. Introduction of the filter-feeding Asian clam *Potamocorbula amurensis* to San Francisco Bay via larvae in ships' ballast water has fundamentally changed the ecology of the Bay. The clam has reached such high densities in this new habitat that it can clear the overlying water of phytoplankton. Since estuaries often support important shellfish fisheries particular problems can also occur if the alien plankton that arrive include harmful algae that can form red tides.

In addition to microalgae and zooplankton, the transport of bacteria in ship ballast water is held responsible for some disease outbreaks in both marine organisms and in humans. For example, the cholera epidemic in Central and South America in the early 1990s is believed to have been due to the initial discharge of *Vibrio cholerae* in ballast water from a ship in Peru. Although *V. cholerae* occurs freely in both salt and freshwater, it preferentially attaches to the chitinous exoskeleton of zooplankton such as copepods. The South American cholera epidemic between 1991 and 1994 resulted in over a million infections and 10,000 deaths.

While it is thought that only a few of the many hundreds of probably introduced species have caused significant problems, in order to try to mitigate the transport of coastal species from one area to another, shipping companies are now encouraged to perform ballast water exchange when at sea. Ballast water exchange involves the flushing of coastal ballast water from the tanks and its replacement with open-ocean water. Since open-ocean species are considered less likely to survive in coastal environments, this practice is thought to be a good safeguard against the spread of alien invaders.

Plankton technologies

While the plankton can create some problems for humans, their vital role in the marine ecosystem should never be forgotten. Plankton may also offer humans some solutions to current technological problems. For example, some of the specialisms among planktonic organisms are even providing inspiration for the science of biomimetics – the field of science that brings biologists and engineers together to copy nature. The most famous example of a biomimetic product is probably Velcro®, which was inspired by the way plant burrs stick to animal fur (and human clothing). It is hoped that the great variety of adaptations to life in the plankton will similarly provide biomimetic inventions. Some of the products that are envisaged include novel adhesives that work in wet environments, nanoscale self-assembling structures in silica, or high-water-content gels based on jellyfish tissue. Perhaps most exciting, however, are the hopes that microalgae may become an alternative carbon source for energy, such as biodiesel. Many microalgae have an oil content of 50% and by producing carbon through harnessing solar energy by photosynthesis they can be used to create a biofuel that is virtually carbon neutral. Large-scale production of microalgae is already under way and it has been estimated that algal biofuels could provide 6% of the global requirement for road transport diesel by 2030.

A vital role

The photic zone is a complex habitat where the diversity of life is huge and the pace of life and death is rapid. Although they are tiny in size, the plankton, through their sheer abundance, play a central role in the ecology of planet Earth. Our understanding of the great variety of organisms in the plankton is improving year by year through the work of scientists using traditional plankton nets and newer technologies to collect data for ever more sophisticated methods of study such as DNA sequence analysis. The photographs on the following pages of this book, which focus on the microalgal phytoplankton and the mesozooplankton – planktonic animals between 0.2 and 20 mm – serve to reveal another of the wonders of the plankton: their astonishing beauty and variety of forms. Photographed at high magnification, these plankton are part of a mysterious world of microscale designs and intricate architectures that provide their owners with novel and beautiful solutions to life in the ocean's photic zone.

PLANKTON CREATED OIL AND GAS Some of the organic carbon in the dead remains of plankton that sank to the sea floor over millenia became buried in the sediments and created oil and gas when it was heated and compressed.

"I proved to day the utility
of a contrivance which will afford me
many hours of amusement and work – it
is a bag four feet deep, made of bunting,
and attached to semicircular bow this by
lines is kept upright, and dragged behind
the vessel – this evening it brought up a
mass of small animals, and tomorrow
I look forward to a
greater harvest".

CHARLES DARWIN, BEAGLE DIARY, 10TH JANUARY 1832
TENERIFFE TO CAPE VERDE ISLANDS

the plankton

Phytoplankton

The phytoplankton – single celled photosynthetic microalgae and photosynthetic bacteria – are at the base of the marine plankton food chain. They are known as primary producers as they use sunlight to convert carbon dioxide into organic carbon compounds (sugars) by the process called photosynthesis – the same process that land plants use to grow. The earliest forms of photosynthetic life, photosynthetic bacteria, are believed to have evolved in the seas around 3.8 to 3.5 thousand million years ago and used hydrogen rather than carbon dioxide in photosynthesis. Around 2.8 thousand million years ago, cyanobacteria containing chlorophyll-*a* appeared that used carbon dioxide for photosynthesis with the by-product of oxygen. The evolution of oxygenic photosynthesis subsequently established an atmosphere rich in oxygen and created the ozone layer setting the conditions for the evolution of more advanced life forms.

Today, the simplest oxygen producing marine photosynthetic prokaryotes in the oceans are still the chlorophyll-*a* containing cyanobacteria, also known as the blue-green algae. Among the eukaryotic microalgae, the diatoms are the largest phytoplankton group and account for about 45% of the ocean's total primary production. While the earliest, reliable fossil evidence dates the origin of diatoms to the Cretaceous period about 145 to 65 million years ago, they are thought to have evolved much earlier during the Triassic, approximately 251 to 199 million years ago. Although some microalgae live as chains of cells or colonies, most exist as single-cells like these *Coscinodiscus concinnus* and they can occur in such vast numbers when they bloom that they can colour the sea surface and are even visible from space.

A bloom of the centric diatom *Coscinodiscus concinnus*

MAGNIFICATION X 140

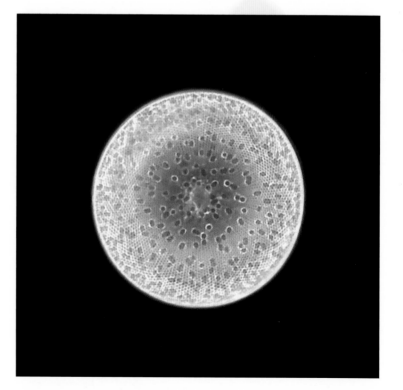

Coscinodiscus concinnus

MAGNIFICATION X 280

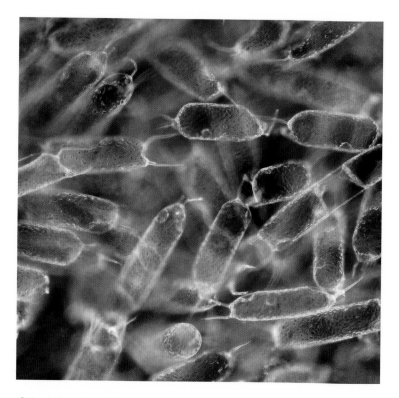

Primary production

The oceans and seas cover more than 70% of the planet and the phytoplankton can occur in such vast numbers that – despite their microscopic size – they account for approximately 50% of all photosynthesis on Earth. Together with land plants, the phytoplankton are thus the principal producers of oxygen (a by-product of photosynthesis). This primary production removes carbon dioxide from the sea and establishes a concentration gradient between the atmosphere and the ocean that leads to a drawdown of this greenhouse gas from the air into the sea surface. Consequently, the phytoplankton are important regulators of the Earth's atmospheric carbon-dioxide content and climate. It has also been suggested that phytoplankton may influence global temperatures by capturing radiation that would otherwise be reflected back to space. Calculations suggest that the warming effect due to phytoplankton may result in a global climate that is up to 0.3°C hotter than if phytoplankton did not inhabit the sea surface.

The chain-forming centric diatom *Stephanopyxis palmeriana*
MAGNIFICATION X 625

Odontella sinensis
MAGNIFICATION X 150

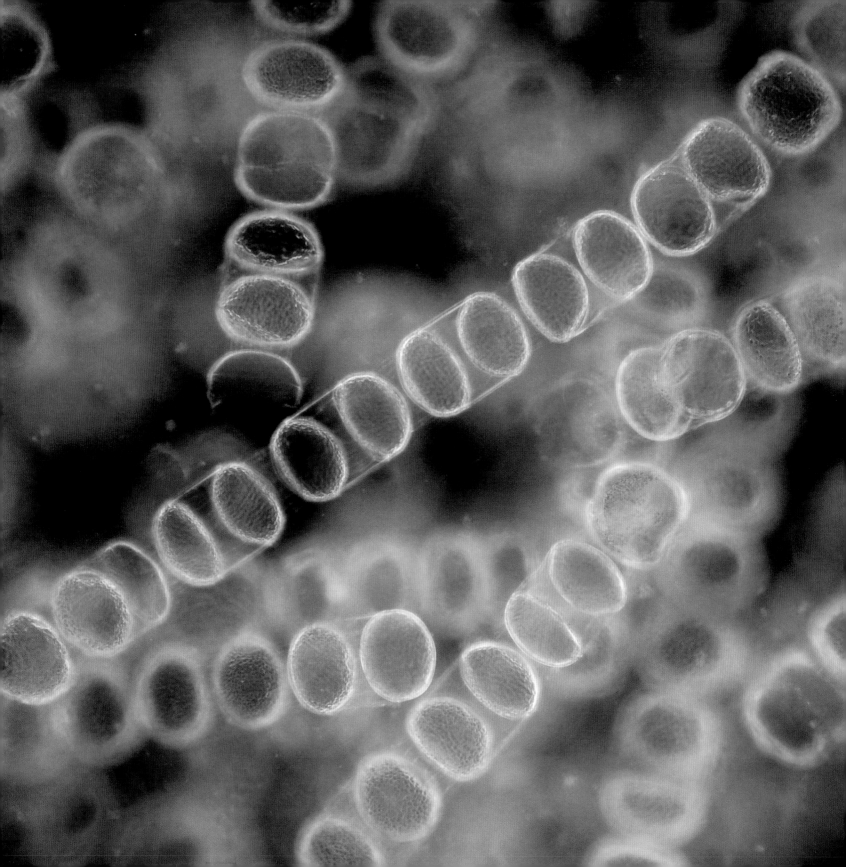

Diatoms

All diatoms have a bi-valved external cell wall – the frustule – made of two silica shells, which makes them heavy, and so they naturally tend to sink. Spines, forming chains of cells, or the presence of intracellular lipids are all strategies they employ to help them float near the surface. If conditions in the plankton turn unfavourable, for example when nutrients like silica become limiting or the temperature becomes too warm, they may reduce their buoyancy or form heavy resting spores and sink through the water column. As soon as conditions become favourable again, such as in the spring, these resting cells are lifted to the surface by vertical mixing as the water warms, to bloom rapidly once again. The steady settling of dead diatoms on the sea bed means they play an important role in the biological carbon pump transferring carbon from the sea surface to the deep ocean. Over millions of years, these settled diatom frustules have formed sedimentary deposits. Today, these sedimentary rocks are referred to as either diatomaceous earth or diatomite and they are mined for use as fine abrasives (often found in metal polish and toothpaste), to make filters for use in water purification, or as a material for thermal insulators.

The pennate diatom *Pleurosigma* sp.

MAGNIFICATION X 430

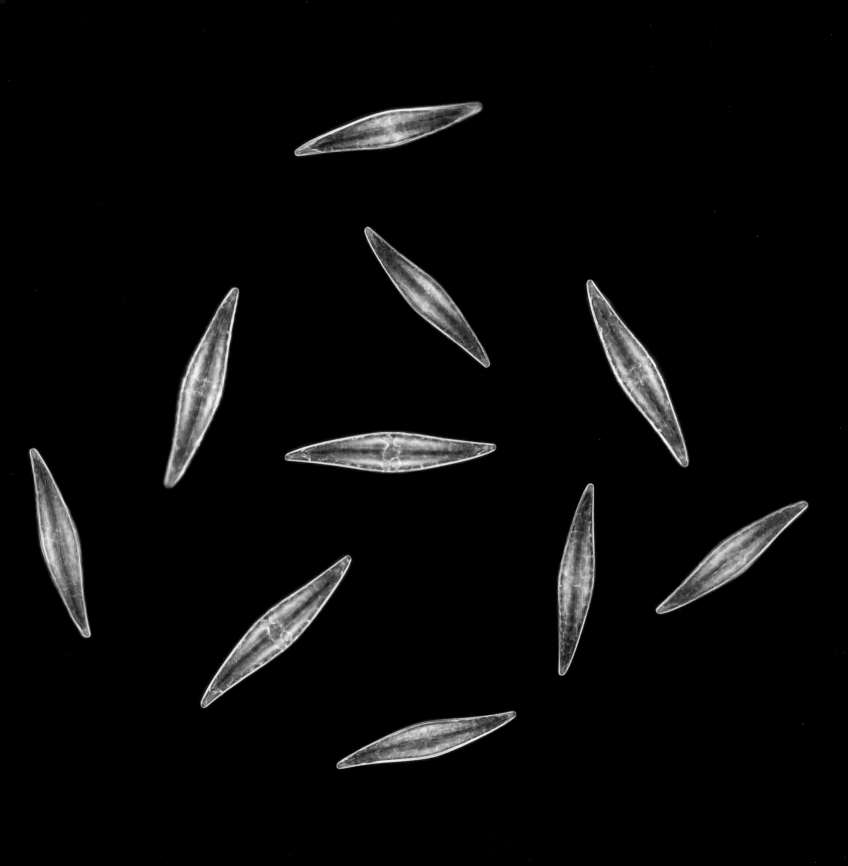

Smaller and smaller

Like *Conscinodiscus* spp., *Rhizosolenia robusta* is a centric diatom, even though it does not look disc-like in shape. Three of the four *R. robusta* cells in the picture have just divided asexually to each make two new diatom cells. When most diatoms divide, each new cell uses one of the two silica valves of the parent and creates another new valve within the parent cell. Consequently, at each cell division the two new cells are approximately one half the size of the parent. Clearly, this reduction in cell size cannot occur indefinitely, and so when a certain minimum size is reached the diatoms produce a cell called an auxospore, which is associated with sexual reproduction. The auxospore lacks a silica cell wall and so the diatom can expand to regain its maximum size. A new silica frustule is then produced and the asexual cycle of cell division recommences.

The centric diatom *Rhizosolenia robusta*

MAGNIFICATION X 150

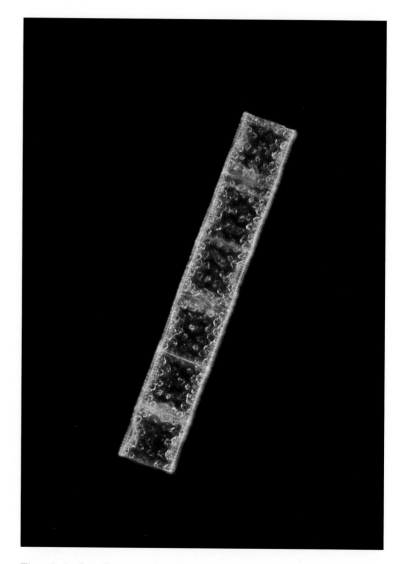

The chain-forming centric diatom *Guinardia flaccida*

MAGNIFICATION X 330

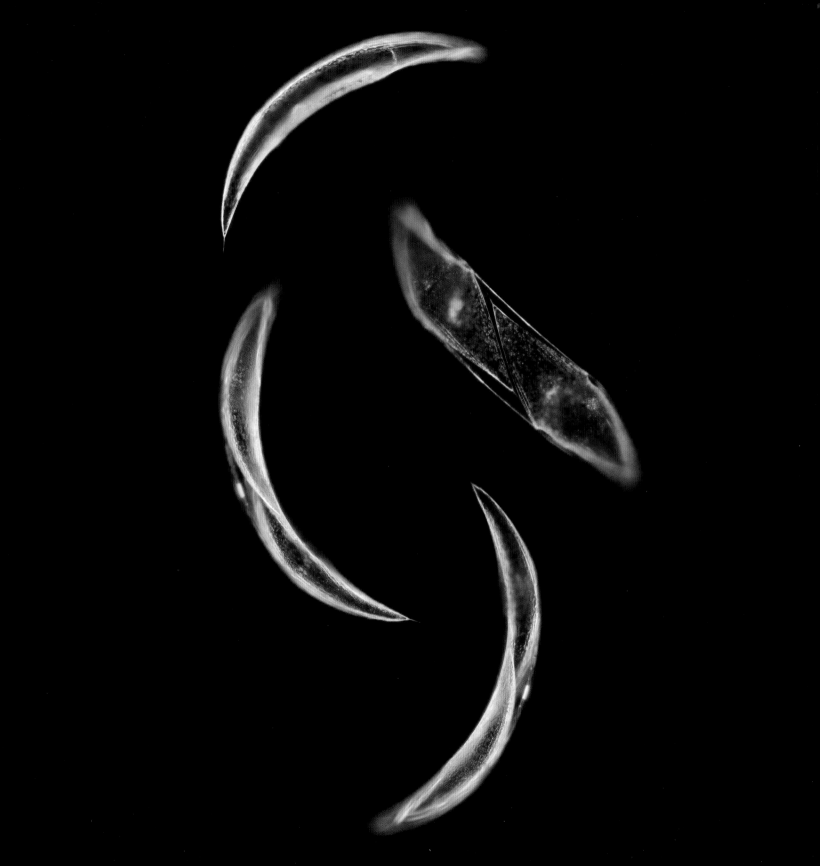

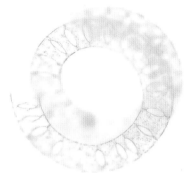

Graceful spirals

This is the chain-forming diatom *Eucampia zodiacus*. This species grows well at low temperatures and it can form an important component of the autumn phytoplankton bloom. In Japanese waters, autumn and winter blooms of *E. zodiacus* have an economic impact on the commercial cultivation of *Porphyra yezoensis*, otherwise know as the seaweed "Nori", which is an important component of sushi. Blooms of *E. zodiacus* deplete the nitrogen in the water column and limit the growth of *P. yezoensis*. The reduction in the production of "Nori" due to this diatom is estimated to cost the Japanese aquaculture industry in excess of a billion yen each year.

The chain-forming centric diatom *Eucampia zodiacus*

MAGNIFICATION X 300

Phytoplankton culture

In 1905, E J Allen began experiments on growing pure algal cultures at the Marine Biological Association on Plymouth Hoe, UK. These early studies developed into the Plymouth Algal Culture Collection under the guidance of Dr Mary Parke. Today, the Collection is a living assemblage of over 450 phytoplankton strains, including 40 type cultures. Now, along with similar facilities in over 65 countries around the world, the Plymouth Collection provides a research centre and offers specialist help on the isolation and culture of different marine phytoplankton.

A culture of the chain-forming diatom *Asterionellopsis glacialis*
MAGNIFICATION X 100

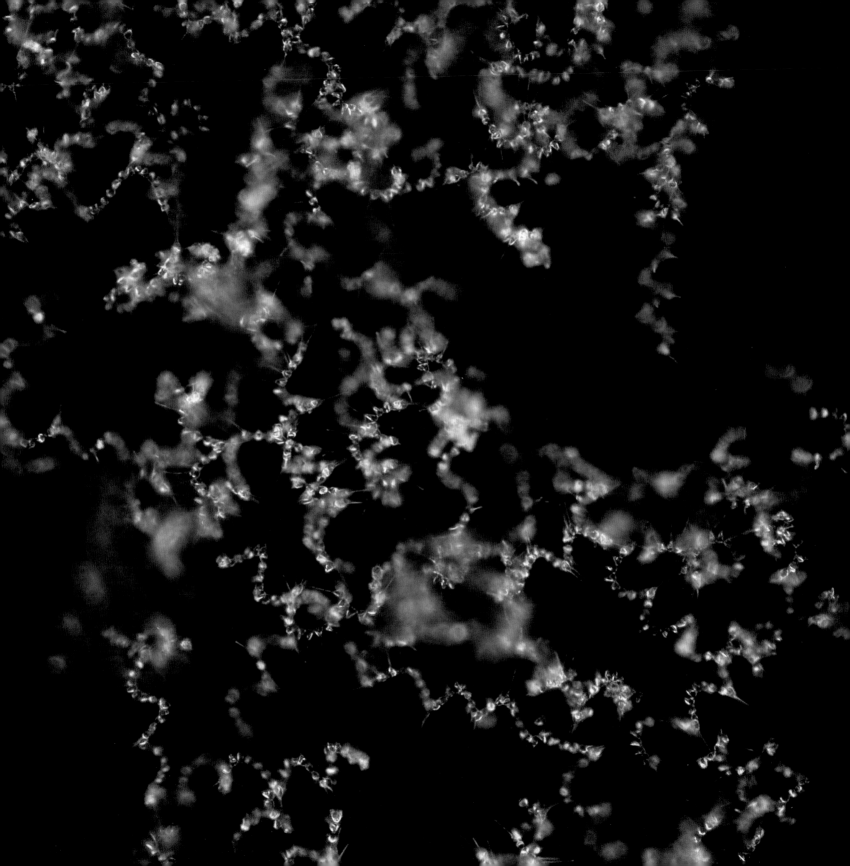

Algal blooms and the smell of the sea

Microalgae can form dense blooms when conditions for growth are good –
suitable light, temperature and nutrients – but when nutrients become
depleted the microalgae will either die or form resting spores. The
prymnesiophyte *Phaeocystis globosa* has large gelatinous, mucus-rich
colonies each comprising hundreds of algal cells. When a bloom of this
species dies it can give rise to accumulations of sulphurous smelling surface
foam that is often blown onto beaches by the wind. The sulphurous smell
is a result of compounds that occur inside many prymnesiophytes and other
microalgae such as dinoflagellates. One of the principal compounds is
dimethylsulphoniopropionate (DMSP). When the microalgae die, the DMSP
they contain is released into the seawater where it is degraded by bacteria and
other phytoplankton into insoluble dimethyl sulphide (DMS). It is the DMS,
along with other algal compounds, that creates the distinctive smell we
associate with the sea.

The prymnesiophyte *Phaeocystis globosa*

MAGNIFICATION X 260

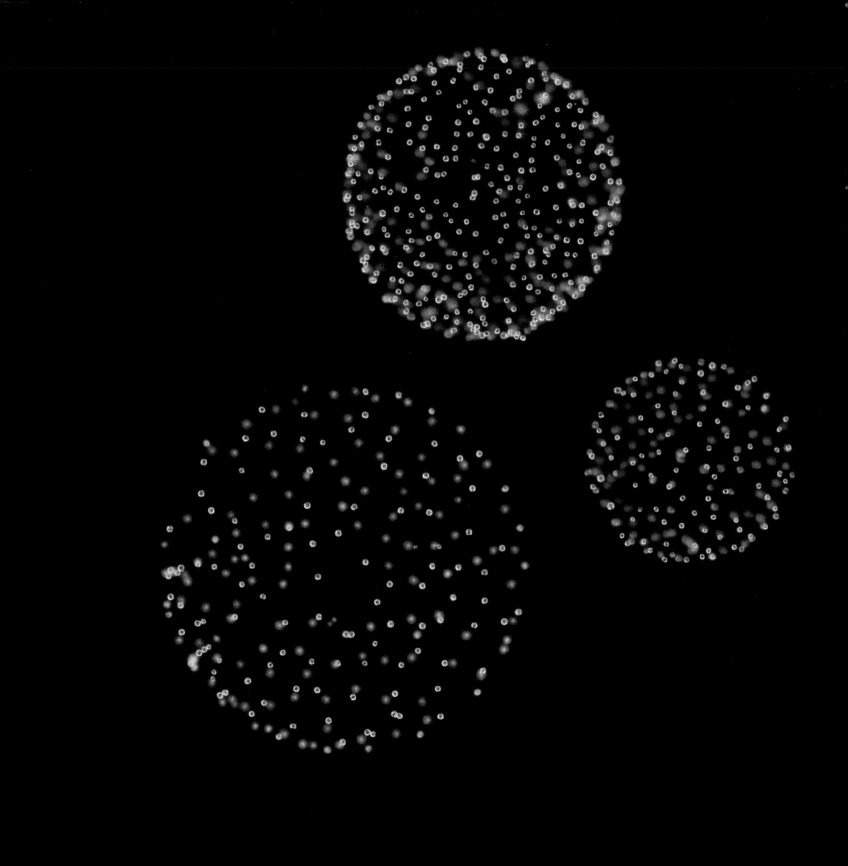

Coccolithophores and chalk cliffs

Coccolithophores are prymnesiophyte microalgae that produce calcite (calcium carbonate) coccoliths inside their cells, which are then excreted onto the cell's surface. Coccoliths are intricately sculpted and although they are usually flat discs, in *Scyphosphaera apsteinii* they are cup-like. Why these microalgae produce coccoliths is not known. However, it is the scattering of light by shed coccoliths that gives the sea the characteristic milky white colour when a large coccolithophore bloom dies. Coccolithophores are important algae in the carbon cycle. Globally, coccolithophores account for the production of over 1.4 billion kilograms of calcite a year helping to sequester carbon dioxide as inorganic carbon. This biogenic production of calcite actually links life in the sea with rock weathering on land through the reaction between carbon dioxide and rock minerals in the terrestrial weathering process. In the case of magnesium silicate-containing rocks, four molecules of carbon dioxide react with the rock mineral to produce magnesium, bicarbonate and silicic acid. After these weathering products are transported by river to the sea they are precipitated biogenically by the plankton. Coccolithophores and amoeboid foraminifera produce calcium carbonate by combining the bicarbonate with calcium in the seawater, while diatoms and radiolarians precipitate the silicic acid as biogenic silica. This entire process results in the long-term net removal of two molecules of carbon dioxide from the atmosphere (although four molecules of carbon dioxide are used in the weathering of magnesium silicate, two molecules of carbon dioxide are given off to the seawater during the biogenic process). Over geological timescales the dead remains of coccolithophores have created massive sedimentary deposits. The 80 m high White Cliffs of Dover on the south coast of England were created almost entirely of single-celled coccolithophores deposited in shallow seas during the Cretaceous period of Earth's history, approximately 145 to 65 million years ago.

The coccolithophore *Scyphosphaera apsteinii*

MAGNIFICATION X 700

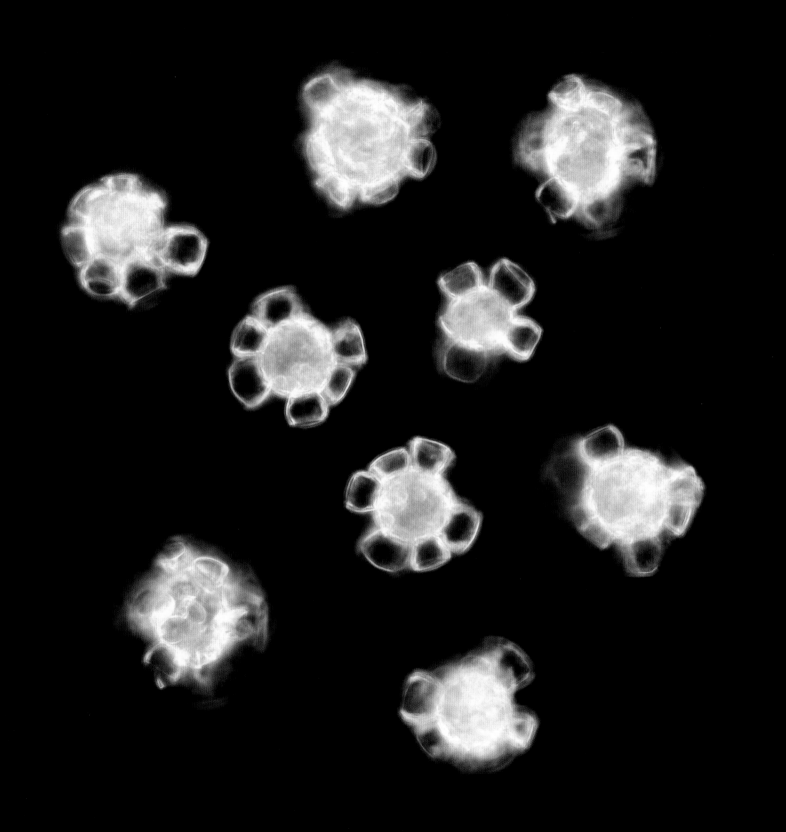

Dinoflagellates

Dinoflagellates are the second largest group of microalgae. Single-celled, they may be either 'naked' or have a cellulose cell wall. Unlike diatoms, which are mainly non-motile (some diatoms such as *Pleurosigma* spp., can move by gliding), most dinoflagellates have two flagella that they use to propel themselves through the water. At sunset many microalgae migrate vertically – by swimming or adjusting their buoyancy – to slightly deeper waters, where nutrients are higher, returning to the surface at sunrise. Many dinoflagellates also have the ability to bioluminesce (emit light) especially when they are grazed by herbivorous zooplankton like copepods and euphausiids. As the intensity of grazing on a dinoflagellate phytoplankton bloom increases, the flashes of light are thought to attract predators of the herbivores and so reduce the grazing pressure. In this way, dinoflagellate bioluminescence is thought to have evolved as a form of plankton 'burglar alarm'.

The dinoflagellates *Ceratium fusus, Ceratium tripos, and Ceratium macroceros* (clockwise from top left)

MAGNIFICATION X 400

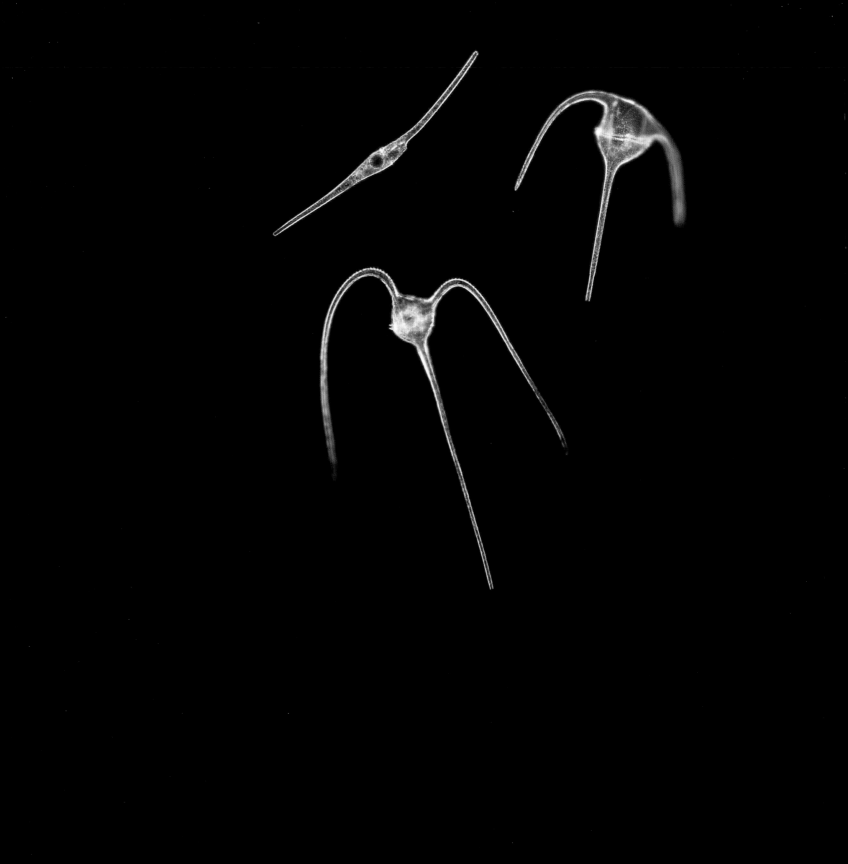

Harmful phytoplankton

While most phytoplankton are beneficial and harmless, some are harmful when they reach high densities. Referred to as harmful algal blooms (HABs) when they occur, some diatom species that have sharp spines may simply cause fish to suffocate by lodging in their gills and causing acute inflammation. Other harmful algae contain toxins in their cells that are poisonous to other creatures that eat them. Humans may even be affected by HABs if we eat animals such as shellfish that have fed upon harmful algae and accumulated the toxins in their tissues. Paralytic shellfish poisoning caused by the algal compound saxitoxin, produced by microalgae like this dinoflagellate *Alexandrium tamarense*, is one example of an effect in humans. Typical symptoms can range from tingling muscles to stomach-ache, sickness, and diarrhoea. In very rare cases it has resulted in death. Because the aquaculture industry (fish and shellfish farming) is especially vulnerable to HABs, many coastal waters are closely monitored to provide an early warning system for their presence so that shellfish are not harvested during a HAB event. As many toxic algae colour the water reddish-brown when they bloom they are often referred to as red tides.

The harmful microalga *Alexandrium tamarense*
MAGNIFICATION X 220

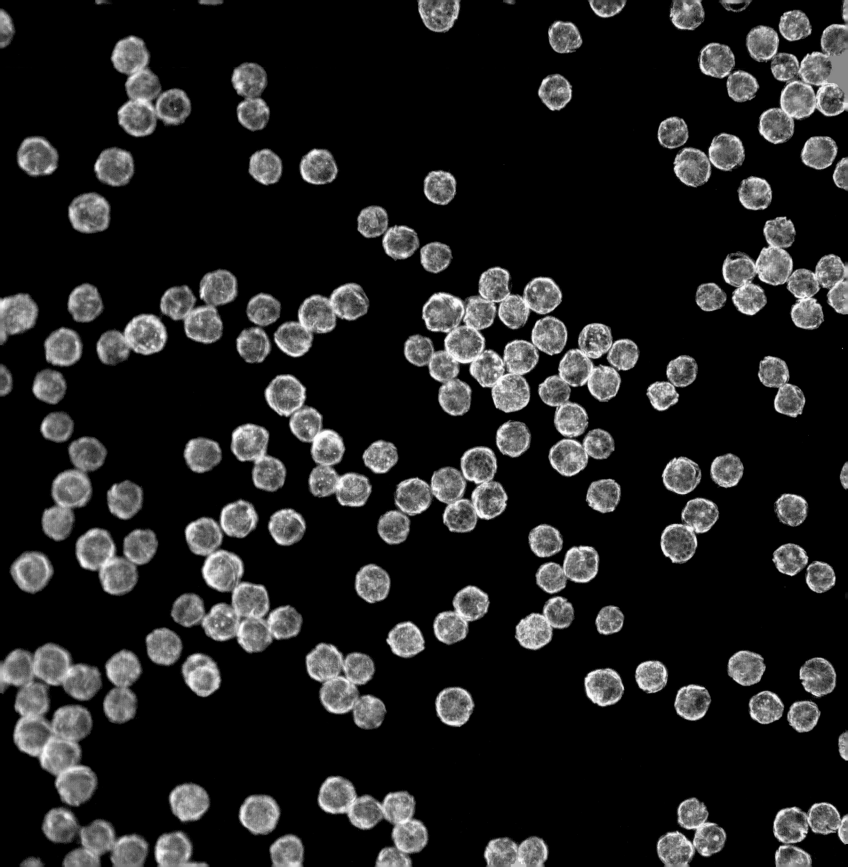

A mixed diet

These cells of the dinoflagellate, *Protoperidinium* sp., are coloured red by oil droplets. Not all microalgae are photosynthetic, some are heterotrophic obtaining their organic carbon by eating bacteria and other microalgae, while others are mixotrophic obtaining their carbon by a combination of photosynthesis and heterotrophy. Consequently, by consuming bacteria, many dinoflagellates are an important component of the microbial loop. This heterotrophic *Protoperidinium* sp., feeds on bacteria, diatoms and other dinoflagellates. The prey, that is first captured using a tow filament, is engulfed by a sheet-like pseudopodium (a temporary extension of the cell's cytoplasm) called the pallium that is produced from between the two antapical spines that can be seen at the base of the lower of these two dinoflagellates. After it is engulfed by the pallium, the prey is then digested externally before the pallium is retracted back into the dinoflagellate; this whole process may take over an hour.

The heterotrophic dinoflagellate *Protoperidinium* sp.

MAGNIFICATION X 500

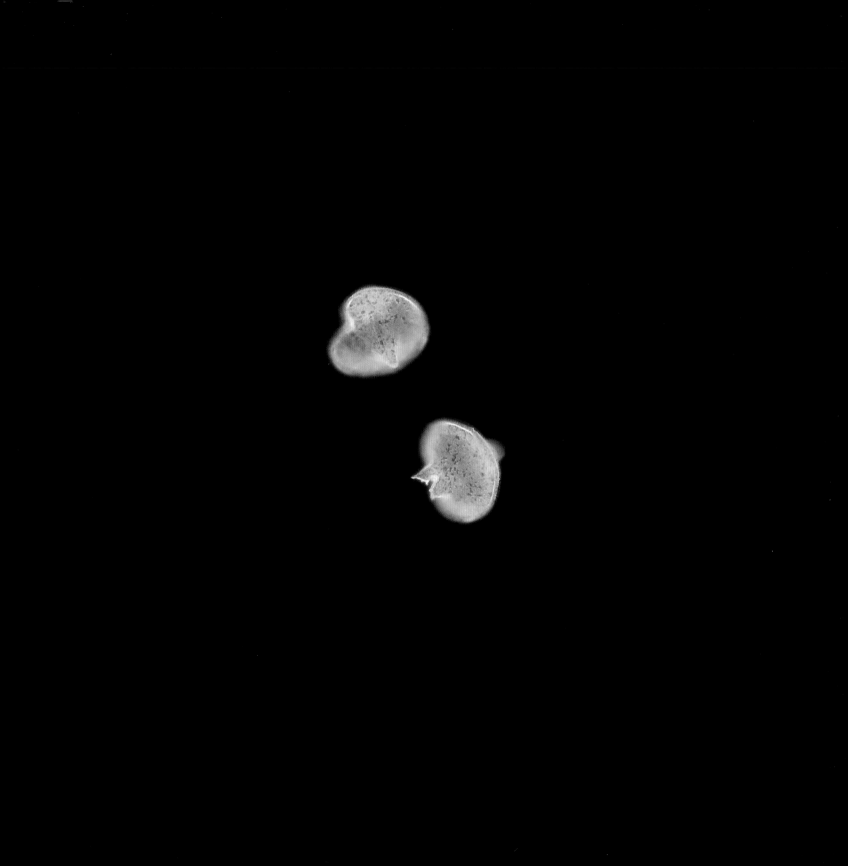

Acantharea and their zooxanthellae

Some microalgae live symbiotically (in association) within other organisms and are known as zooxanthellae. These single-celled protists (animal-like cells) are sarcodine Acantharea and they each contain golden-brown zooxanthellae inside their cell. The zooxanthellae provide the Acantharea with a carbon food source from their photosynthesis and in return the microalgae gain nutrients from their host. Although acanthareans can be very abundant in the plankton, little is known about their ecology as they are fragile and difficult to sample. The needle-like skeletal rods radiating from the centre of each acantharean are made of celestite (strontium sulphate). When a bloom of acanthareans dies the spines fall through the water column to the sea floor and in this way acanthareans can affect the levels of strontium in the surface water. Acantharea belong to a larger group of organisms known as the Sarcomastigophora, which includes the silicate producing Radiolaria and the Foraminifera. Like all sarcodines, Acantharea are active predators in the plankton, feeding on bacteria, microalgae and small zooplankton.

Single-celled Acantharea **and their zooxanthellae**

MAGNIFICATION X 400

Zooplankton

Phytoplankton are the food source of the secondary consumers, including the mesozooplankton – animals between 0.2 and 20 mm. By feeding on phytoplankton and also on bacterioplankton, the zooplankton are a critical link in transferring energy through the food chain to higher trophic levels, such as fish, seabirds and whales. Scientists subdivide the mesozooplankton into two groups, the holozooplankton and the merozooplankton. The holozooplankton are those animals that spend their whole life in the plankton, from egg to larva to adult, such as copepods. The merozooplankton are those animals that only complete some of their life cycle in the plankton, such as crabs and barnacles that have planktonic larvae.

A merozooplankton, a zoea larva of the angular crab *Goneplax rhomboides*

MAGNIFICATION X 100

A holozooplankton, the copepod *Subeucalanus crassus*

MAGNIFICATION X 25

Holozooplankton

Copepods dominate the mesozooplankton biomass and calanoid copepods are among the largest. While many copepods graze on phytoplankton, some species are predatory and feed on other zooplankton. Many copepods undergo daily and seasonal vertical migrations from the surface into much deeper waters. Several species spend the daylight hours at depth to avoid predation and only rise to the surface at sunset to feed during the night. Interestingly, some of the largest diel vertical migrations are seen among coloured copepods that are perhaps most visible during daylight hours. The 4 mm-long copepod, *Pleuromamma robusta* descends to over 400 m daily by swimming and adjusting its buoyancy; this represents a journey of over 100,000 times its own body length and would represent a daily round trip of about 400 km in human terms. Copepods only have primitive eyes. The two long antennae arising from the head of this *Calanus helgolandicus* help it sense its environment.

The calanoid copepod *Calanus helgolandicus*
MAGNIFICATION X 90

The calanoid copepod *Temora longicornis*
MAGNIFICATION X 25

Plankton biodiversity

Scientists still do not know the full extent of biodiversity in the plankton, how many species there are, because their small size makes them difficult to identify. For example, female *Calanus helgolandicus* copepods can only be distinguished from their sister species, *C. finmarchicus*, by the amount of curvature on the inner border of the basipodites and the number and size of the teeth on the fifth swimming leg – the fifth pereiopod. The border of the pereiopod basipodites in *C. helgolandicus* is more curved with fewer, larger teeth compared to *C. finmarchicus*. One relatively new approach to understanding biodiversity in the plankton is to study their DNA – their genetic code. In a recent global study of 20 oceanic copepod species that had been described on the basis of their morphology (body shape) an analysis of variation in their DNA sequences uncovered four new species that would otherwise have remained unknown.

The fifth pereiopod and basipodites of *Calanus helgolandicus*

MAGNIFICATION X 170 (INSET X 340)

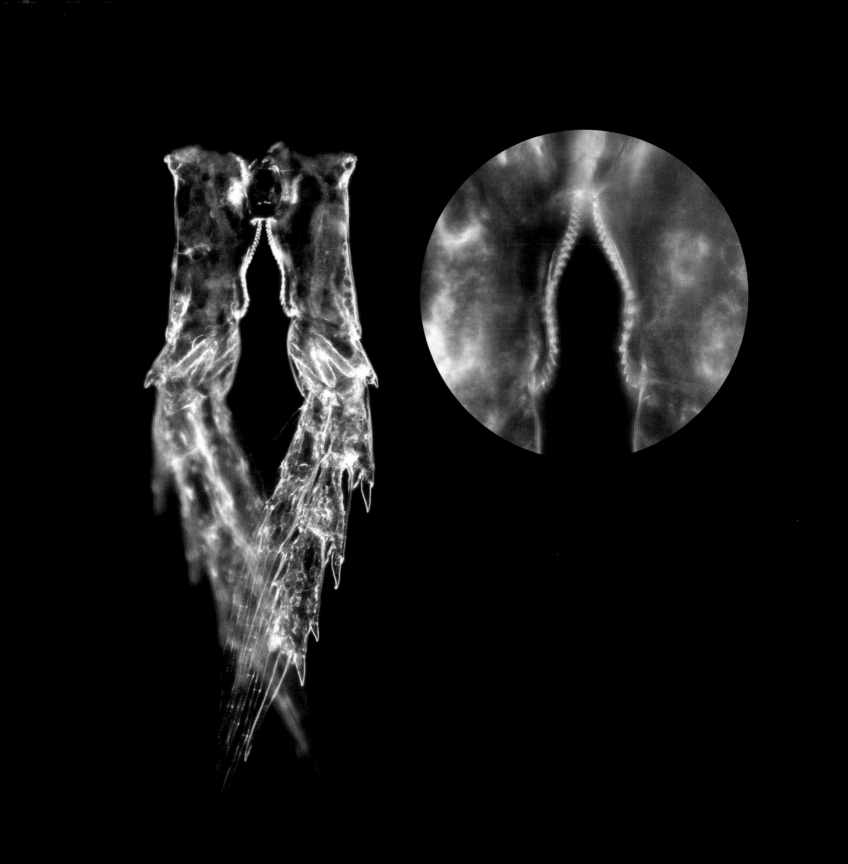

Copepod aggregations

Seasonally, many copepods reduce their metabolic rate – diapause – and overwinter at great depth. Along with euphausiids, copepods are the food source of some of the largest creatures on Earth, the baleen whales. The Arctic bowhead whale, *Balaena mysticetus*, can grow up to 100 tons in weight and 18 m long and their mouths contain 300 baleen plates each measuring 300-450 cm long that they use to filter tens of thousands of copepods from every mouthful of water. The whales feed mainly during the winter months on the dense aggregations of copepods that form in deep water when the copepods diapause.

A group of calanoid copepods *Calanus* spp.

MAGNIFICATION X 15

Escaping predation

To help avoid being eaten by a predator many plankton have evolved
defence mechanisms. These may include body armour such as spiny
extensions or bristles, flashes of light, or escape strategies such as the
vigorous escape 'jump' of the copepod *Acartia tonsa*. The presence of a
predator may be detected in the water by mechanical vibrations, by its
scent or by visual means. Copepods have a range of predators from other
zooplankton such as amphipods, chaetognaths and jellyfish, to fish, whales
and seabirds. In *Acartia tonsa* a rapid fall in light intensity, such as would
occur due to the shadow cast by a predator, causes the copepod to flick its
swimming legs vigorously. These vigorous thrusts give an escape speed of
between 400 and 600 mm per second and each 'jump' displaces the 1.5
mm adult copepod about 10–15 mm in the water. While these jumps may
not help escape the gaping mouth of a whale they may be sufficient to
escape other predatory zooplankton and fish larvae.

The copepod *Acartia clausi*

MAGNIFICATION X 100

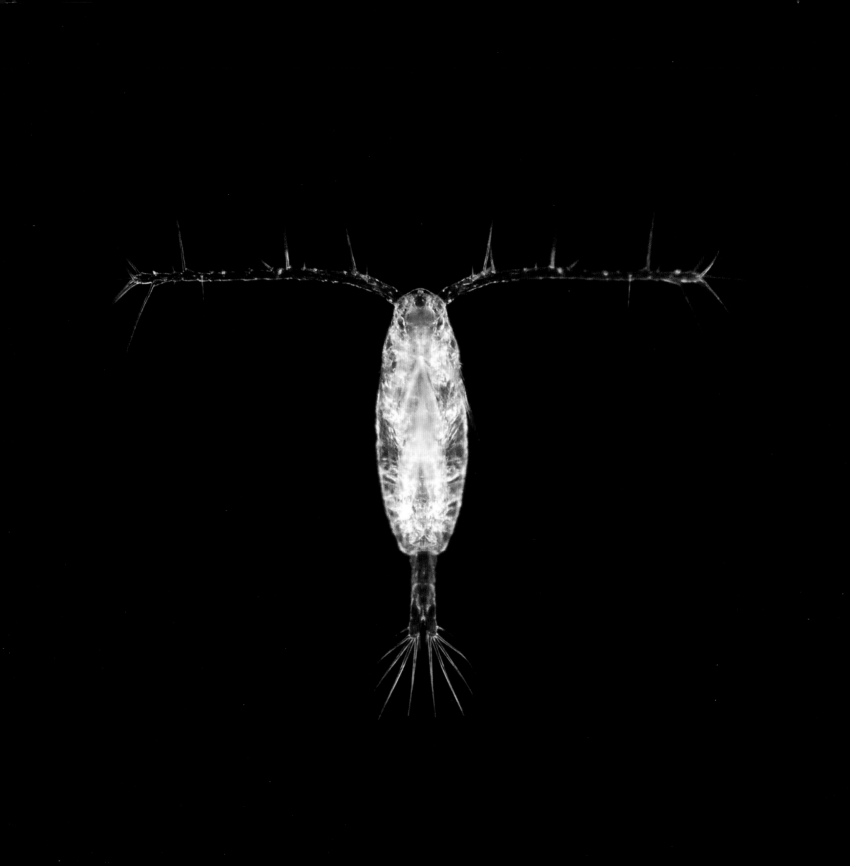

Finding a mate

In order to reproduce, male and female copepods must first find
each other amongst all the other plankton. With only simple eyes, male
copepods locate a mate by using their antennae, first to follow a scent
trail in the water and then to explore the surface of the potential partner
to detect species-specific, surface-borne scents called pheromones. Once
a mate is found, many male copepods then have a means of stopping
the female from escaping. In *Centropages typicus*, one of the male's
hind limbs, the fifth pereiopod, is modified into a large pincer-like chela
(claw) that it uses like a handcuff to grasp and hold the female. Accurate
positioning of the male and female during mating is essential since
successful copulation relies on the male's spermatophore (sperm-
containing sac) being placed precisely on the female's urosome (tail).
Depending on the species, the female copepod will either release her eggs
directly into the water or will carry them with her in an external egg sac.

A female *Centropages typicus*
MAGNIFICATION X 75

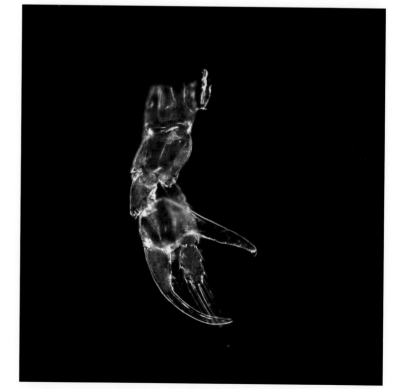

The handcuff-like claw of a male *Centropages typicus*
MAGNIFICATION X 400

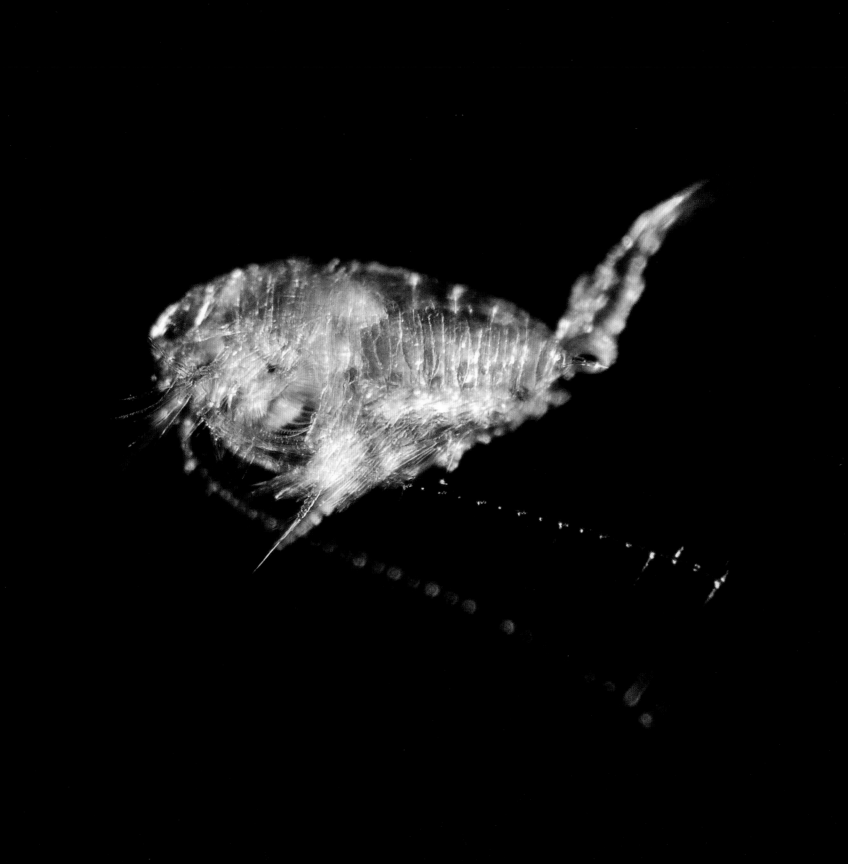

Copepod reproduction

While adult copepods can reproduce all year round in many waters,
especially in the tropics, spring is a stimulus to reproduction in temperate
regions. Many copepods produce large numbers of small eggs that they
release freely into the water. However, only some of these eggs will survive,
as many will be eaten by other zooplankton, including other copepods,
often even cannibalistically. An alternative strategy used by many
copepods, like this cyclopoid (top) and harpacticoid (bottom), is to
produce fewer, larger eggs and give them some protection by carrying
them in an egg sac until the juveniles – known as nauplii – hatch. By
remaining attached to the parent these eggs will migrate with the adult at
sunrise to the relative safety of deeper waters during the day where
predation is reduced.

The copepods *Oithona similis* (top) and
Euterpina acutifrons (bottom)

MAGNIFICATION X 80

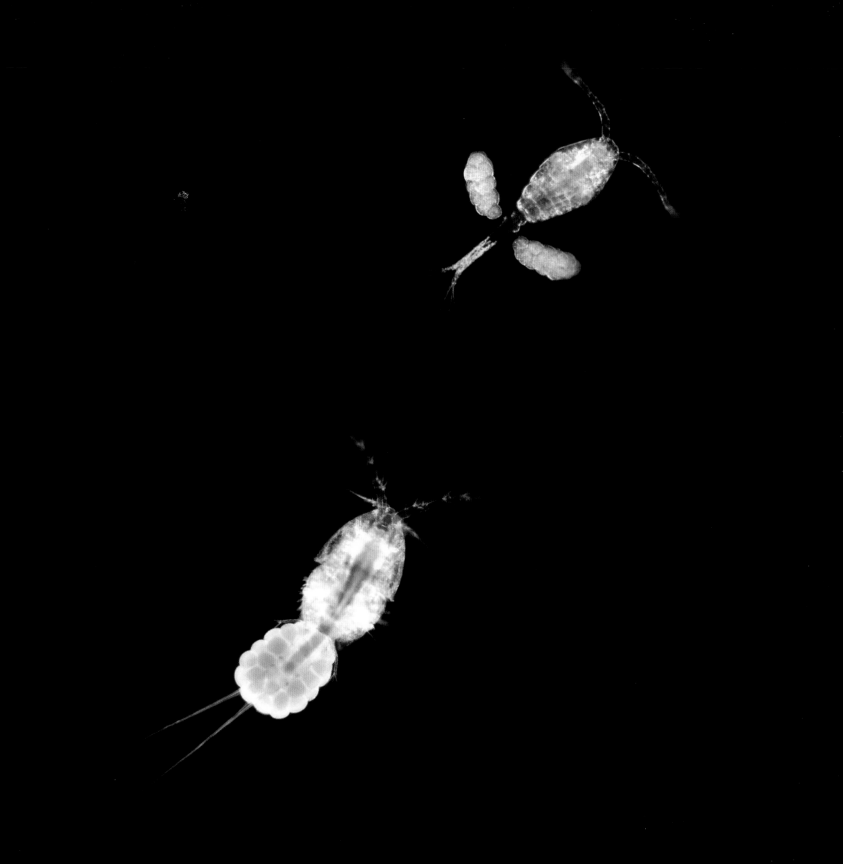

The copepod life cycle

The copepod egg first hatches into a nauplius larva. As the young copepod grows it passes through several naupliar stages acquiring more limbs at each moult. The last naupliar stage is followed by six copepodite stages that each become more adult-like with more limbs and segmentation. At the onset of winter, the fifth copepodite of some species descends in the water column, in some cases to depths of 2,000 m or more, where it enters the dormant physiological state called diapause. Other species may overwinter as resting eggs. Next spring, the copepodites awaken, perhaps triggered by rising sea temperatures or increasing day-length, and they rise to the surface to arrive in time for the start of the spring phytoplankton bloom. At this point they moult into the last copepodite stage to become an adult copepod that will find a mate and reproduce to start the life cycle again.

A copepod nauplius larva

MAGNIFICATION X 800

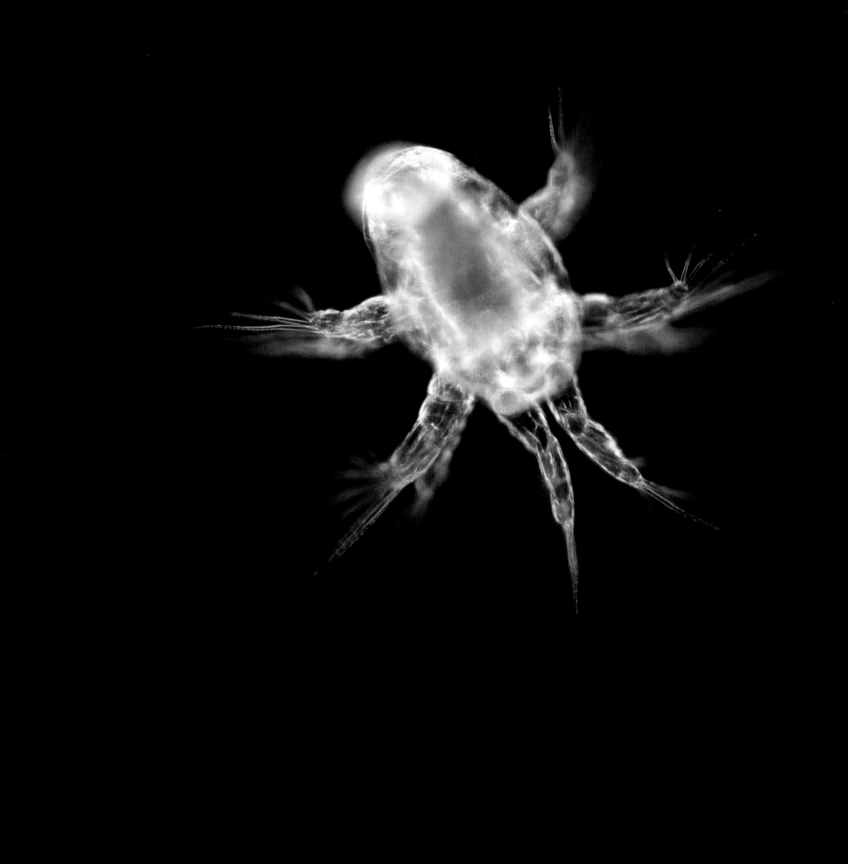

Sea lice

This parasitic caligid copepod lives on the skin or gills of fish. The adult lays eggs into the water that hatch into a juvenile nauplius stage that eventually moults into a copepodite. The free-living juvenile copepodite then locates a fish and attaches to its host first by using its clawed antennae and then by an adhesive. Once attached, the copepodite moults into the adult stage that is known as the chalimus. Otherwise known as sea lice, caligid copepods often cause open wounds that can lead to infection and to the death of the host fish. Because of their harmful effects these parasitic copepods are often a significant economic nuisance to fish farms.

A parasitic caligid copepod

MAGNIFICATION X 80

Euphausiids

Important members of the holozooplankton, most euphausiids are filter-feeders using their foremost limbs – the thoracopods – to 'comb' microalgae from the water. On their abdomens, euphausiids have five pairs of limbs called swimmerets – the pleopods – that, as their name suggests, are used for swimming. Most euphausiids also have specialised organs on their sides called photophores that are bioluminescent however, the true purpose of these organs is still unknown. Perhaps the best-known euphausiid is the Antarctic krill, *Euphausia superba*. Like most euphausiids, Antartic krill swarm in vast numbers and are a key plankton species in the food chain where they are eaten by fish, seals, whales, albatross and penguins. A commercial krill fishery also exists in the Southern Ocean where krill are caught for processing into either animal feed or for human consumption as okiami. The western Antarctic Peninsula in the Southern Ocean is one of the seas warming fastest in response to global climate change. During winter the juvenile krill survive by feeding upon sea-ice microalgae and so their food supply is determined by the geographical extent of the sea ice. Winter sea ice extent has declined as the Antarctic has warmed and this has led to a two-fold decrease in the abundance of krill, affecting the pelagic food chain.

The North Atlantic euphausiid *Nyctiphanes couchii*

MAGNIFICATION X 70

Mysids

The pereiopods (thoracic limbs) of mysids each have two branches, although the fifth pair can be reduced in some species, and they are used for swimming and to catch prey. Most mysids are omnivores feeding on both phytoplankton and zooplankton, switching their diet depending upon what is most abundant in the water at the time. In turn, many marine animals eat mysids. While many mysids are epibenthic (living on the sea bed) some species are holoplanktonic and, wherever they live, they can occur in such large swarms that they can control the abundance of other species by the effects of their predation. Within the three lobe-like uropods at the end of the abdomen of this mysid is a statocyst. The statocyst is a small sac that contains a mineral deposit – the statolith – that is surrounded by sensory hairs. The statolith serves to indicate orientation and aid balance.

A mysid

MAGNIFICATION X 50

Human relatives

The larval stage of a doliolid possesses a notochord (a flexible rod-like structure of supporting cells) and so, like humans, doliolids belong to the phylum Chordata. Adults alternate between colonial, asexual and solitary, sexual generations. The eight bands of muscle around this barrel-shaped adult sexual stage help the animal swim. Contractions of the annular muscles draw water in through the buccal siphon at the top and expel it as a jet of water from the atrial siphon at the bottom. Beating cilia (hair-like cells) on the internal gill slits also draw in a current of water. A mucus sheet behind the gills traps food particles from this stream of water before it leaves the body.

The doliolid *Doliolum nationalis*

MAGNIFICATION X 100

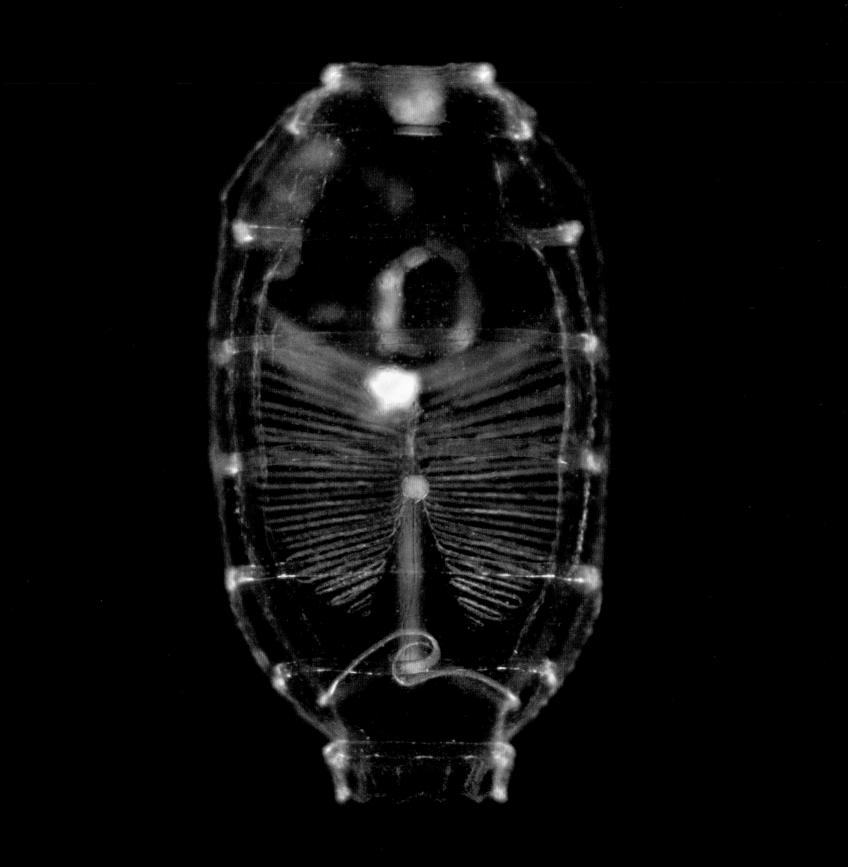

Water fleas

These cladocera (water fleas) use their two large antennae like oars to propel themselves through the water to help catch phytoplankton and copepod nauplii. Water fleas have an unusual reproduction since they show cyclical parthenogenesis. For most of the growing season each female will reproduce asexually (without the need for fertilisation by a male) by brooding successive generations of diploid eggs within her carapace that will all hatch into females. Consequently, for most of the year, the entire population in the plankton is female. In autumn, at the end of the growing season, the females start to produce male offspring and haploid eggs and they reproduce sexually. Now, the females lay eggs into the water instead of giving birth to live offspring. The fertilised eggs then sink to the sea bed where they remain dormant in the sediments until the spring, when they are triggered to hatch into embryos. The parthenogenetic stage of the water flea's life cycle then starts again.

The water flea *Podon leuckartii*

MAGNIFICATION X 100

The water flea *Evadne nordmanni*

MAGNIFICATION X 90

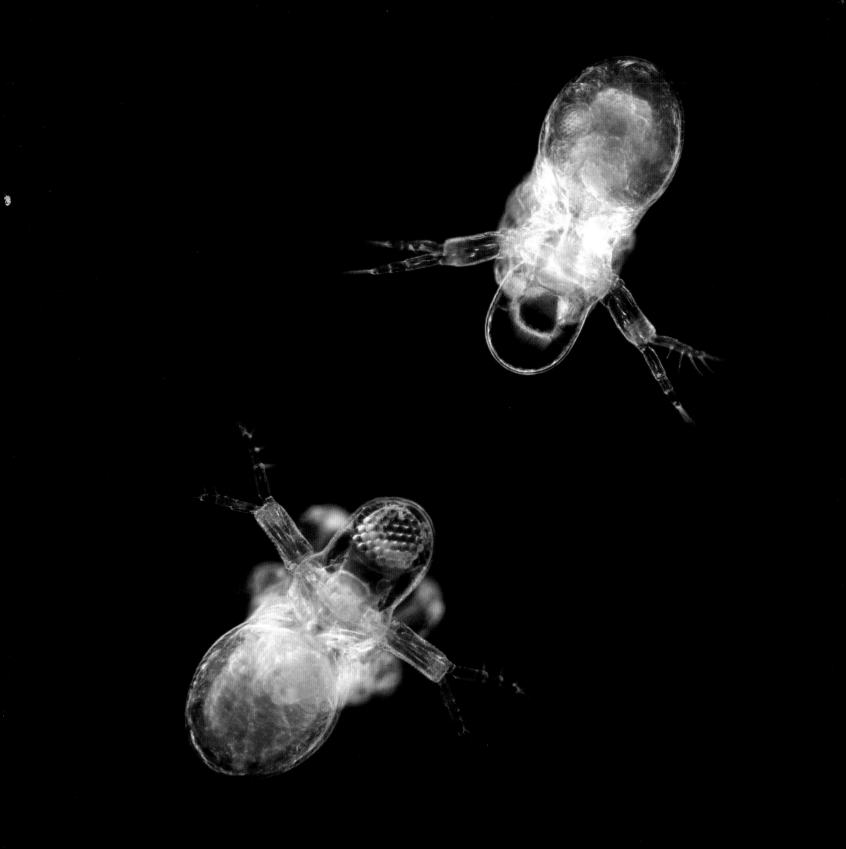

Hyperiids

Most hyperiid amphipods, such as *Hyperia galba*, are parasites of jellyfish and salps, and they live attached to the surface of the animal on whose tissue they feed. This hyperiid, however, is one of the few free-living species in the zooplankton. Free-living hyperiids are voracious zooplankton predators using their large eyes to hunt for copepods, chaetognaths, euphausiids, pteropods and fish larvae. The free-living amphipod *Themisto gaudichaudii* can occur in such vast swarms in the Southern Ocean that it is often the most abundant predator of other zooplankton. In turn, it is an important food for fish and seabirds. Because hyperiid amphipods can occur in such large numbers they can control mesozooplankton biomass through the effects of their predation. Consequently, hyperiids can exert an important influence on the downward flux of carbon to the sea bed.

A free-living hyperiid amphipod

MAGNIFICATION X 65

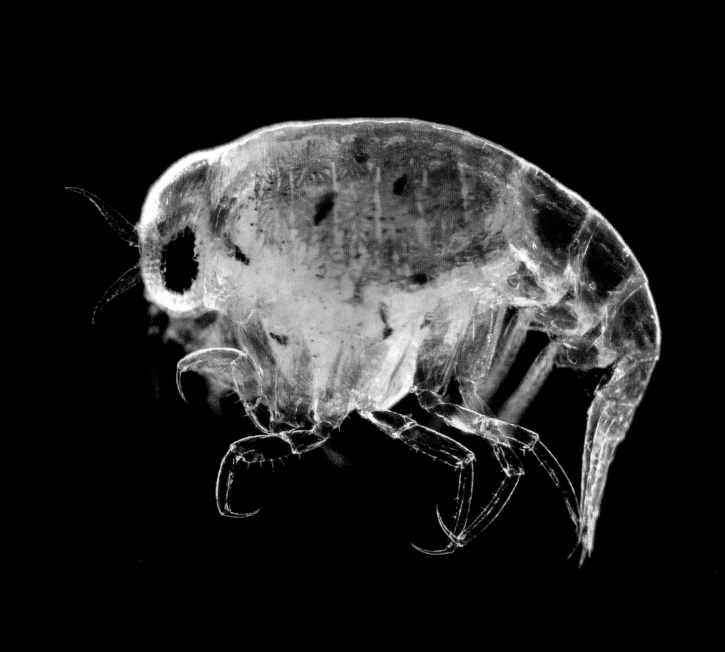

Sea butterflies

Formerly called pteropods, and still referred to by this name, these snails belong to the Thecosomata. The snails swim by continually flapping the two wing-like extensions of their foot, hence they are commonly known as sea butterflies. The snail *Limacina retroversa*, like other pteropods, is an active predator feeding on both phytoplankton and small zooplankton. To catch their prey they produce a mucus web in the water that is many times their own size. This web sieves plankton from the water and when it is full of food it is drawn back to the mouth to be eaten. Pteropods occur throughout the world's oceans and like many zooplankton they tend to spend the daylight hours at depth and migrate to the surface at night. These snails can be very abundant and they are an important food of many pelagic fish such as mackerel and herring. In the Southern Ocean, the pteropods *L. retroversa australis* and *L. helicina antarctica* have been recorded at densities of up to 800 and 1400 individuals per cubic metre. Occurring in such large numbers, pteropods can be as important as krill in the Southern Ocean food web. Currently, scientists are concerned that rising levels of carbon dioxide in the atmosphere may be harming the growth of many planktonic organisms that have calcium carbonate shells or skeletons. When carbon dioxide dissolves in seawater it forms a weak acid and, as levels of this gas are increasing in the atmosphere, the sea surface is becoming more acidic. Pteropod shells are made of aragonite, a form of calcium carbonate that dissolves readily as the pH of seawater decreases (the acidity increases). Because of the importance of pteropods in the Antarctic food web there are worries that the ecology of the Southern Ocean may be particularly vulnerable to the effects of ocean acidification.

The North Atlantic pteropod *Limacina retroversa*

MAGNIFICATION X 60

Sea angels

These planktonic, predatory sea slugs are molluscs belonging to the Gymnosomata and can be found from the Arctic to the Antarctic. Together with their relatives the shelled sea butterflies, such as *Limacina retroversa*, they were known previously as pteropod molluscs. However, unlike the sea butterflies, which have a shell throughout their life, sea angels only have a shell when they are embryos. While the juvenile *Clione limacina* pictured here are just 5 mm long, the adults can reach up to 5 cm. Sea angels swim by flapping the two wing-like parapodia that are derived from the animal's foot. Although sea angels normally swim quite slowly, they are active predators in the plankton where they feed upon sea butterflies and other zooplankton, and they can show rapid bursts of speed to catch their prey.

The sea angel *Clione limacina*

MAGNIFICATION X 20

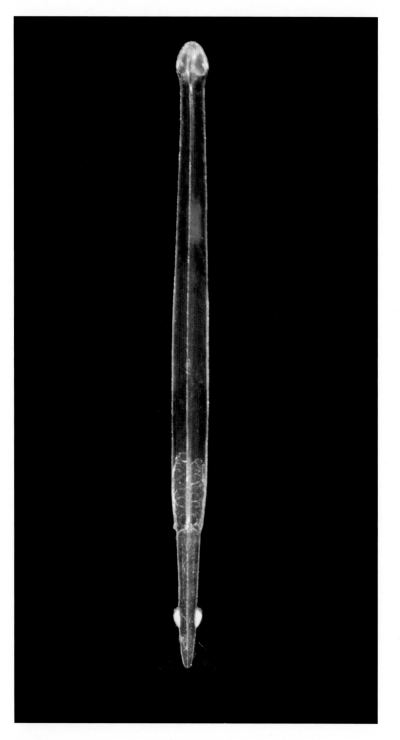

Arrow worms

The likely fate of most zooplankton is to be eaten. Planktonic chaetognaths, also known as arrow worms because of their slender appearance, can occur in large numbers and are important predators of other zooplankton and fish larvae. Chaetognaths swim with a undulating motion using their tail fin for propulsion. The body is almost transparent and the head bears a pair of simple eyes. The name 'chaetognath' comes from the Greek words *chaite* and *gnathos* meaning 'bristle jaws'. The bristle-like hooks used to seize prey are clearly visible on the head of this animal. Like many planktonic predators, chaetognaths exert both top-down and bottom-up control of fish recruitment by feeding on fish larvae and the planktonic food of fish larvae.

The head of the chaetognath *Sagitta setosa*

MAGNIFICATION X 650

Sagitta setosa

MAGNIFICATION X 17

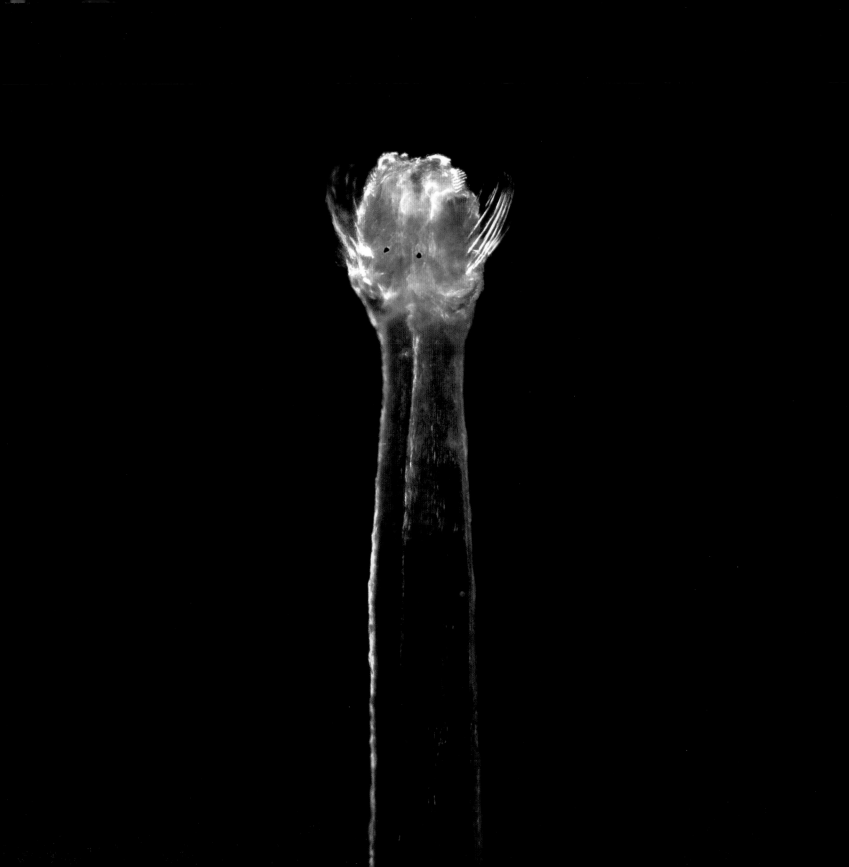

Planktonic paddlers

Relatively little is known about the biology of the fast swimming and almost transparent polychaete worm *Tomopteris helgolandica*. It swims with a wriggling, snake-like motion by using the two lobes of its divided limb-like parapodia as paddles, hence its common name of 'paddle worm'. When disturbed, *T. helgolandica* will roll up into a ball and sink in the water. Despite lacking teeth, the worm is an active predator of zooplankton, especially chaetognaths and fish larvae. The head bears a pair of eyes equipped with lenses, a pair of short antennae and two long 'tentacles' that may extend up to two-thirds of the body length; these tentacles are modified parapodia, each supported by a stiff rod made of chitin called the aciculum. A remarkable feature of this polychaete is that it is bioluminescent. Between the lobes of the parapodia is a light-producing organ that emits a brief flash of yellow light when the animal is disturbed and which may act as a defence mechanism to startle predators.

The paddle worm *Tomopteris helgolandica*

MAGNIFICATION X 12

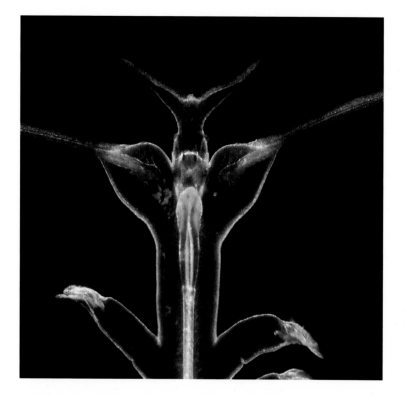

Head of *Tomopteris helgolandica* showing its red eyes

MAGNIFICATION X 20

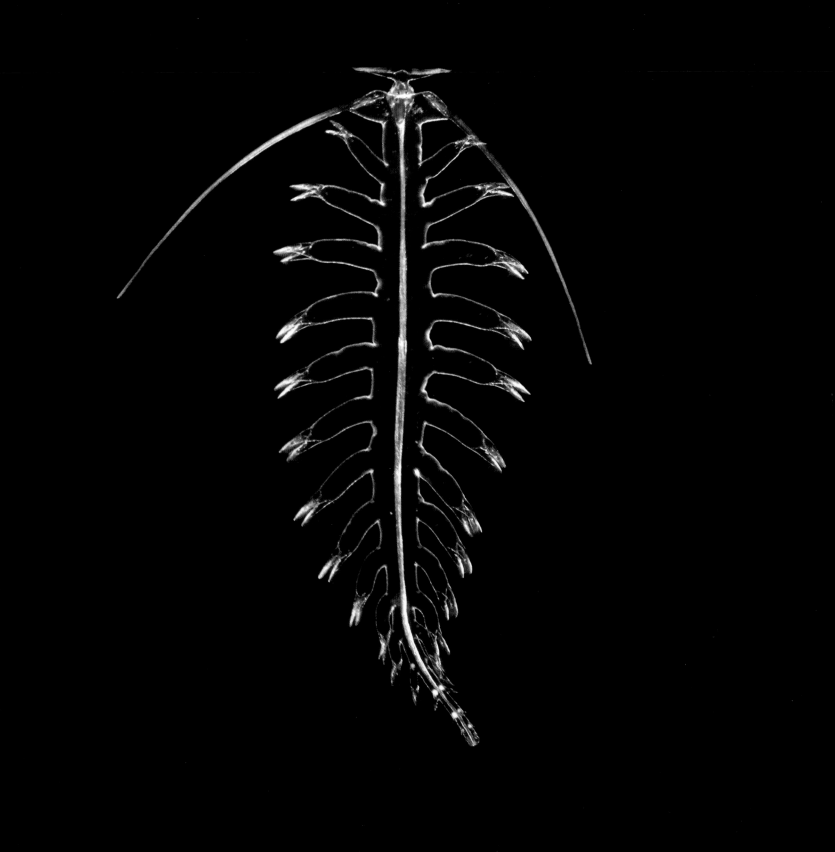

Siphonophores

Belonging to the Cnidaria (jellyfish) these are the transparent bell-like nectophores of the colonial siphonophores, *Muggiaea atlantica* (top) and *Muggiaea kochi* (bottom). These two species of siphonophore are distinguished by the length of their tube-like somatocyst that can be seen along the inside of the nectophore. In *M. atlantica* the somatocyst extends the full length of the nectophore, whereas it only extends for half the length in *M. kochi*. The gold-coloured drop of oil that helps the colony float can be seen near the apex of the somatocyst of *M. atlantica*. Each individual in the adult colony performs a unique function. The bell-like nectophore pulsates and is used for swimming and trails behind it three types of cell, the gastrozooids used in feeding, reproductive gonozooids and stinging dactylozooids used for defence and catching food. Siphonophores are major predators in the plankton and they can reproduce asexually by budding to create large swarms of up to 500 individuals per cubic metre. Siphonophores feed on other zooplankton such as copepods and fish larvae and so, like many other predatory zooplankton, they may influence the survival of juvenile fish.

The nectophores of *Muggiaea atlantica and Muggiaea kochi*
(top and bottom)

MAGNIFICATION X 50

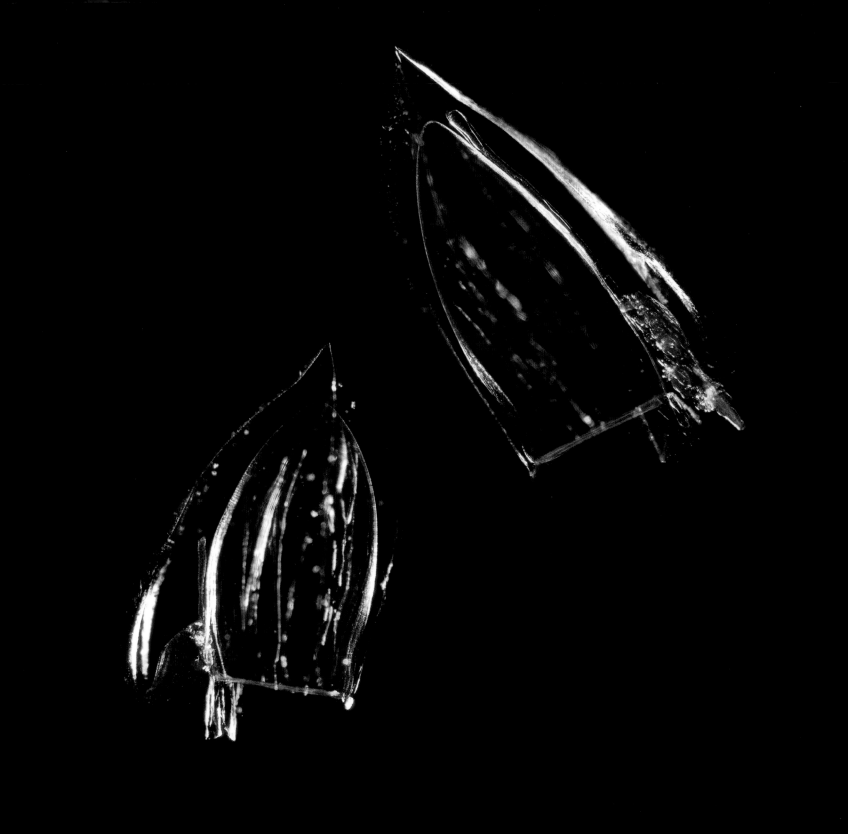

Comb jellies

Ctenophores – also known as comb jellies or sea gooseberries – are important predators in the plankton. While still at the mercy of the ocean currents, most ctenophores have eight rows of fused cilia on their surface that beat synchronously to propel them through the water. Looking like rows of combs, it is these hair-like cells that give the animals their name since 'ctenophore' comes from the Greek words *ctena* (comb) and *phora* (bearer). Ctenophores are carnivores and most possess a pair of retractable tentacles that are covered with sticky cells called colloblasts that they use to catch zooplankton, fish eggs and fish larvae. When the eastern Atlantic ctenophore *Mnemiopsis leidyi* was introduced accidentally via ships' ballast water into the Black Sea in the 1980s it had no natural predators and became very abundant. Within ten years the Black Sea fishery had collapsed as *M. leidyi* out-competed the fish larvae for their zooplankton food. Fortunately, the recent introduction of another ctenophore, *Beroe ovata*, which is a predator of *M. leidyi*, now appears to be controlling their abundance. In the meantime, however, *M. leidyi* has spread from the Black Sea to the Caspian and Mediterranean Seas, and most recently into the North Sea and the Baltic where there are new concerns for the effect it may have on these important fisheries.

The ctenophore *Pleurobrachia pileus*
(the sticky tentacles are retracted into their pouches in the top two individuals)

MAGNIFICATION X 10

By-the-wind sailors

A few marine organisms occupy the air–water interface, half in and half out of the sea, and are known as 'pleuston'. A striking example of such a creature is the predatory jellyfish *Velella velella*, commonly known as the by-the-wind sailor, shown here (from top to bottom) in a side-view, a bird's-eye view and a fish's-eye view. On its upper surface the animal has a stiff, vertical vane made of chitin that runs diagonally from side to side acting as a small sail to catch the wind and propel it across the ocean's surface. In some *V. velella* the vane, which makes the animal sail at 45 degrees to the wind, is angled to the right and in others it is angled to the left; the prevailing winds sort these two forms so that a particular variant dominates on opposite sides of the Atlantic and Pacific Oceans. Below the vane is an oval float that comprises a series of concentric, air-filled, chitinous tubes to provide buoyancy. The olive-brown colour of the fringe of tissue that surrounds the float is due to the presence of zooxanthellae. The zooxanthellae are single-celled, photosynthetic, dinoflagellate microalgae and it is thought that they provide the *V. velella* with carbon compounds from their photosynthesis. In return, the jellyfish gives the zooxanthellae some protection, nutrients and, by floating on the surface, access to plenty of light. Hanging in the water beneath the *V. velella* extend two types of polyps surrounding a central mouth called the gastrozooid. Tentacle-like polyps called dactylozooids, which bear stinging cells called nematocysts are used to catch copepods, fish eggs and other small plankton that come within reach. Shorter, reproductive polyps called gonozooids bud male and female sexual medusae into the water.

The by-the-wind sailor *Velella velella*

MAGNIFICATION X 10

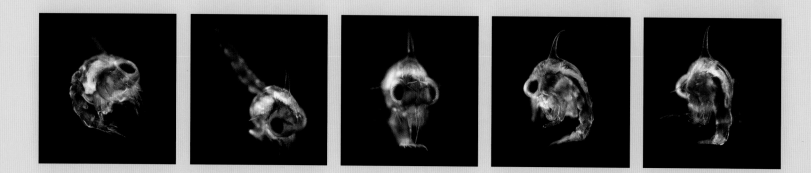

Merozooplankton

The merozooplankton are animals that complete only part of their life cycle in the plankton and they include the larval stages of many benthic organisms (creatures that live on the sea bed), such as worms, crabs, barnacles, starfish and mussels. The word plankton comes from the Greek word *Planktos*, which means 'wanderer' or 'drifter', and a planktonic larval stage is regarded as a way to enable slow-moving or sessile benthic species to disperse to new locations. The merozooplankton can make up to 50% of the total zooplankton biomass and their different adaptations for feeding, flotation, locomotion and defence against predators, create a variety of weird and wonderful forms.

The zoea larva of the common spider crab *Maja squinado*

MAGNIFICATION X 250

Adaptations to life

The first larval stage of a crab is called the zoea. Young zoea larvae first feed on phytoplankton and then on small zooplankton. Most crab larvae are relatively easy to recognise due to characteristic dorsal and lateral spines in brachyuran crabs (true crabs, e.g. spider crab larvae), or the possession of a long rostrum and backward extensions of the carapace in anomuran crabs (e.g. porcelain crabs). Shown here is the larva of the porcelain crab, *Pisidia longicornis*, which, as an adult, lives on the seashore under rocks. The spiny extensions of many zoea are thought to serve two purposes. Firstly, they may act as a deterrent against being eaten and, secondly, they increase the surface area to volume ratio and therefore may help the animal float. By helping flotation the spines enable the larva to expend less energy swimming to stay near the surface and so more of the energy from its food can contribute towards growth. Growing rapidly and minimising the time spent in the plankton is important since the planktonic phase is a time of high mortality in the life cycle of these species.

The javelin-like zoea larva of the porcelain crab *Pisidia longicornis*

MAGNIFICATION X 160

Crab larvae

The legs that this crab zoea uses for swimming can be seen clearly, as can the abdomen with its forked telson and the thoracic spines that serve to reduce predation and aid flotation. In the adult crab the abdomen is tucked away under the body and, in the female, it is used to hold the developing eggs. This larval thumbnail crab is unrecognisable from the adult that lives in the mud on the sea bed. Indeed, there are still planktonic larvae for which scientists are unsure what adult animal they will become. One approach to help determine this is to rear the larvae to adulthood in the laboratory. However, this can be a painstaking process especially as the specific requirements for larval growth are often unknown. A new shortcut to identification is offered by DNA sequence analysis. By comparing DNA sequences obtained from adult and larval forms it is possible to match a larva with the adult animal it becomes. One global project, called the Census of Marine Life (CoML), is aiming to obtain a DNA sequence from all marine organisms.

The zoea larva of the thumbnail crab *Thia scutellata*

MAGNIFICATION X 100

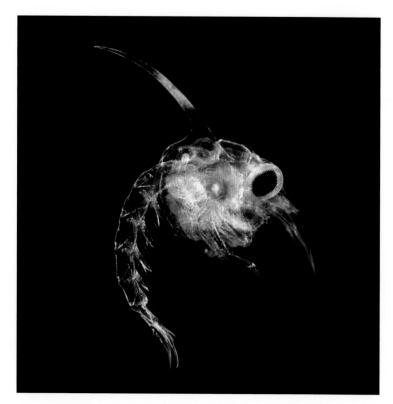

A merozooplankton, a zoea larva of a swimming crab

MAGNIFICATION X 50

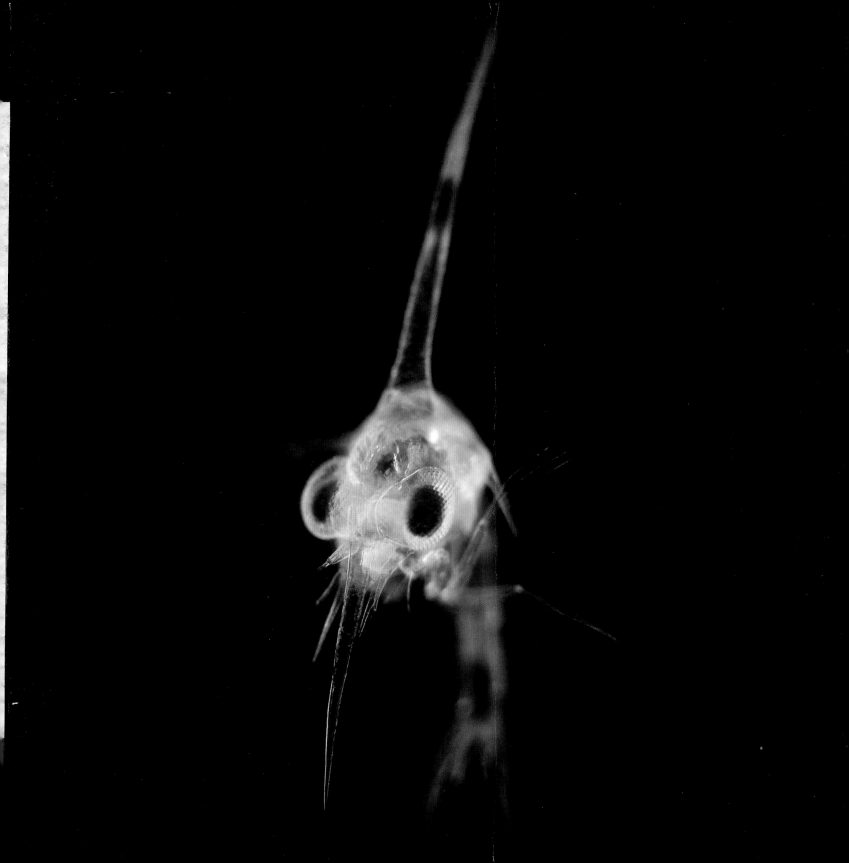

Life lower down

Unlike most crab zoea larvae, these larvae of *Ebalia* sp. have no
dorsal spine on the carapace and only rudimentary lateral spines.
This *Ebalia* zoea lives further below the surface and so it is thought
that long spines, which are considered to help flotation in other
species, are unnecessary. It should be noted, however, that many
of the assumptions scientists make about larval 'adaptations' are
speculative as it is difficult to determine their true purpose. The
larvae of *Ebalia* spp., have the habit of rolling up into a tight ball
when disturbed, which may be a defence mechanism in the
absence of long spines.

The spineless zoea larva of *Ebalia* sp.

MAGNIFICATION X 150

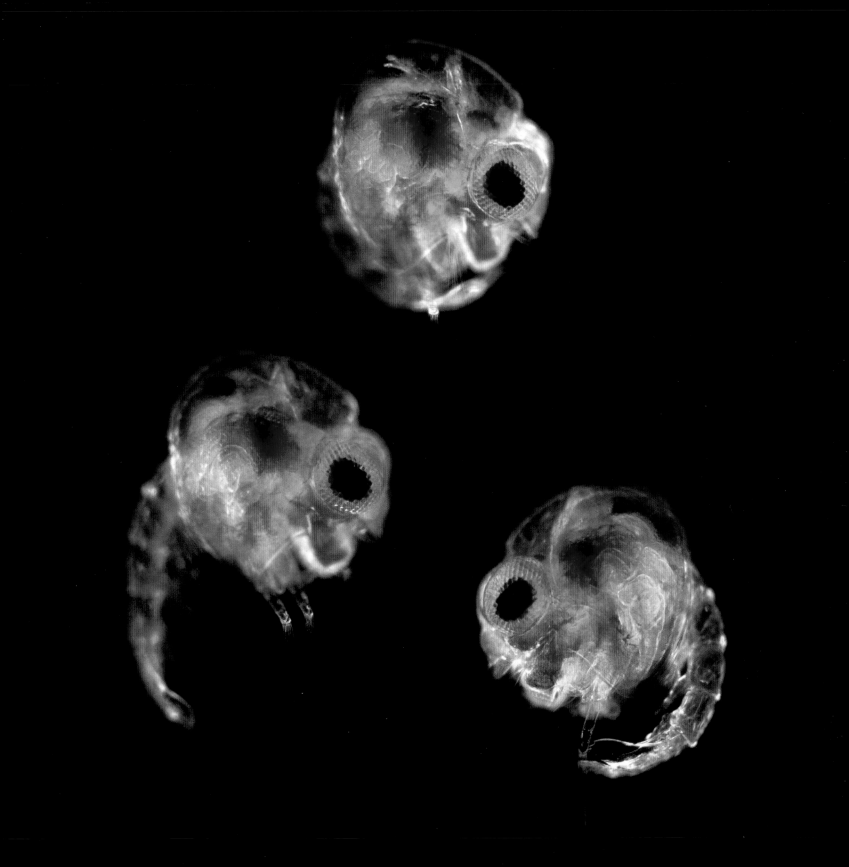

Voracious predators

After feeding and growing in the plankton, the early-stage crab zoea larva metamorphoses into the megalopa. Looking much more crab-like and with a pair of claws, the megalopa stage is short-lived and feeds voraciously on other zooplankton. The megalopa is flattened dorso-ventrally, unlike the almost spherical zoea, and after about a week it sinks to the sea bed where it moults into a juvenile crab. Crabs are important predators on the sea bed and can structure communities with particular effects on the recruitment of bivalves, such as the mussel, and juvenile flatfish.

A crab megalopa larva

MAGNIFICATION X 90

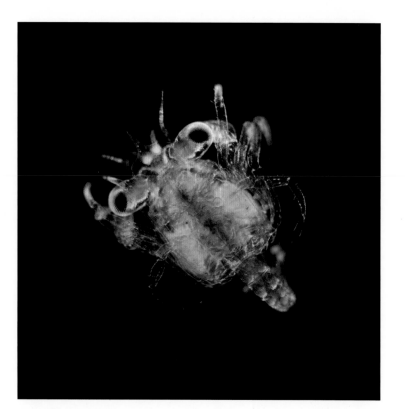

A crab megalopa larva

MAGNIFICATION X 40

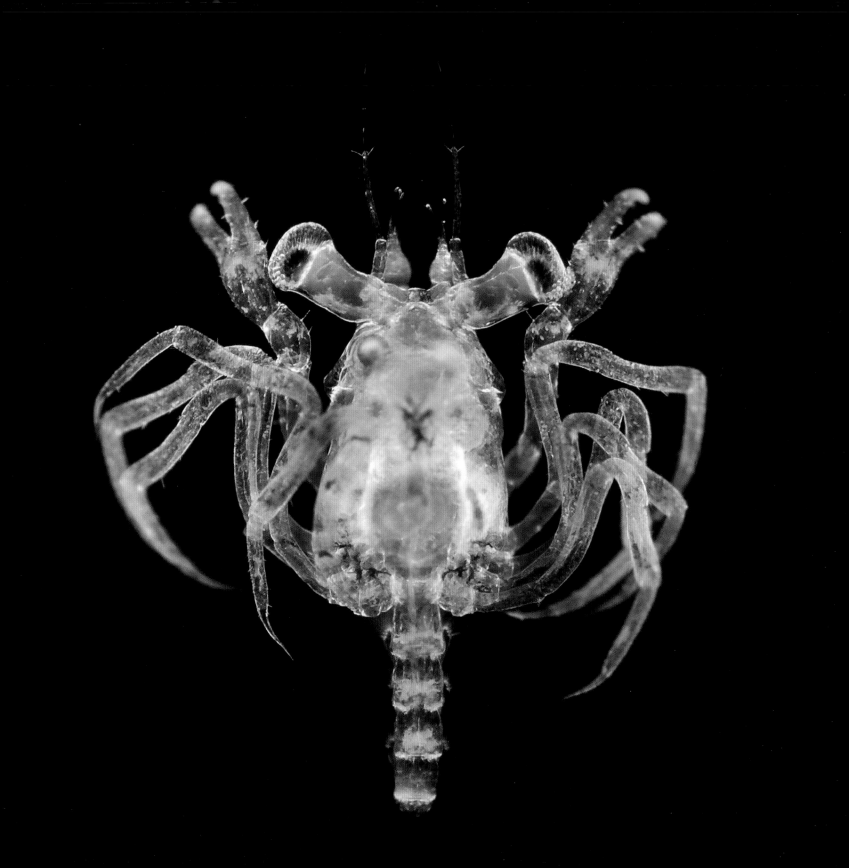

The plankton year

Events in the plankton occur in succession. At temperate latitudes the plankton year begins with the spring phytoplankton bloom triggered by an increase in day length and temperature. Almost immediately, overwintering copepod eggs hatch and the diapause of overwintering juvenile copepods ends. Benthic organisms like shrimps also begin to reproduce and fish spawn so that their larvae can take advantage of the increasing primary and secondary production in the plankton. In the Northern and Southern Hemispheres, the plankton year is brought to a close in the autumn as nutrients become depleted and day length and temperature decrease.

The planktonic larva of a shrimp

MAGNIFICATION X 250

Fisheries

Fisheries include more than catching fish like the cod, *Gadus morhua*,
for a fish-and-chip supper. In coastal areas fisheries also include what
are collectively known as shellfish: shrimps, prawns, crabs, lobsters, and
molluscs, such as scallops, clams and mussels. Shellfish fisheries provide
an average total world landing of around 13 million tonnes each year and
they rely upon the survival of the planktonic larval stages of these animals.
Only a tiny fraction of the total number of larvae produced each year will
survive the planktonic phase to settle and become adults on the sea bed.

The larva of the shrimp *Processa edulis*

MAGNIFICATION X 150

Vision

Being able to see is important for capturing prey and avoiding predators. This crab zoea larva has a compound eye made up of many cells, similar to the eye of an insect. The setae that arise from the tips of the limbs (seen to the left of the eye) are also sensory, providing information on vibrations in the water. Other plankton, such as copepods, arrow worms and barnacle nauplii, possess more simple eyes that can detect only light and motion. The copepod *Corycaeus anglicus* supplements its simple light sensor with two spherical lenses, one on each side of the top of its head. These lenses direct light down separate cones to the light receptor that lies near the centre of its body. It is thought that these simple eyes help *C. anglicus* avoid predators and detect the movement of small jellyfish, such as the medusae of *Obelia* spp., upon which it feeds.

The compound eye and sensory setae of a swimming crab zoea larva

MAGNIFICATION X 330

The copepod *Corycaeus anglicus* showing one of the paired, spherical lenses on its head.

MAGNIFICATION X 90

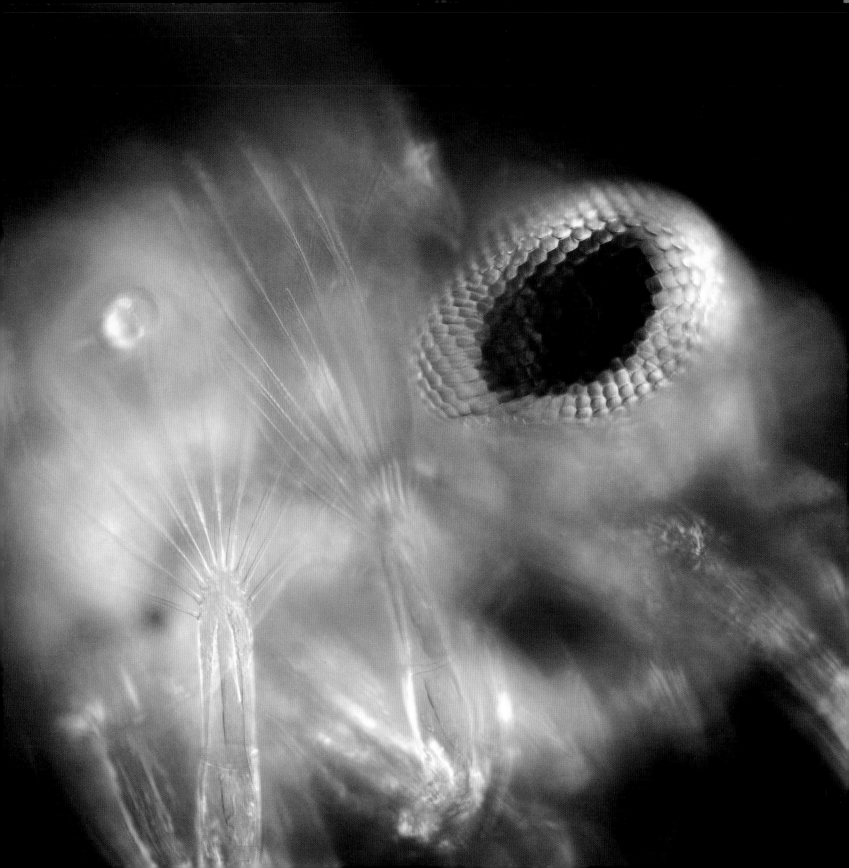

Insect relatives

Crabs and shrimps belong to the phylum Arthropoda, which also includes the insects. These creatures share a basic body plan, a feature that was recognised by the scientist Robert Hooke FRS (1635–1703) who was a contemporary of Christopher Wren and Isaac Newton. Hooke made a series of spectacular drawings of insects, sponges, bryozoans and other planktonic creatures that he published in 1665 in his book, *Micrographia*. On page 178, in an experiment that would now be considered cruel to repeat, Hooke noted the similarity between the compound eyes of crabs, lobsters and shrimps and those of insects, writing:

"Thirdly, that thofe which they call the eyes of Crabs, Lobfters, Shrimps, and the like, and are really fo, are Hemifpher'd, almoft in the fame manner as thefe of Flies are. And that they really are fo, I have very often try'd, by cutting off thefe little movable knobs, and putting the creature again into the water, that it would fwim to and fro, and move up and down as well as before, but would often hit itfelf againft the rocks or ftones; and though I put my hand juft before its head, it would not at all ftart or fly back till I touch'd it, whereas whil'ft thofe were remaining, it would ftart back, and avoid my hand or a ftick at a good diftance before it touch'd it. And if in cruftaceous Sea-animals, then it feems very probable alfo, that thefe knobs are the eyes in cruftaceous Infects, which are alfo of the fame kind, onely in a higher and more active Element"

The compound eye of the zoea larva of *Maja squinado*

MAGNIFICATION X 250

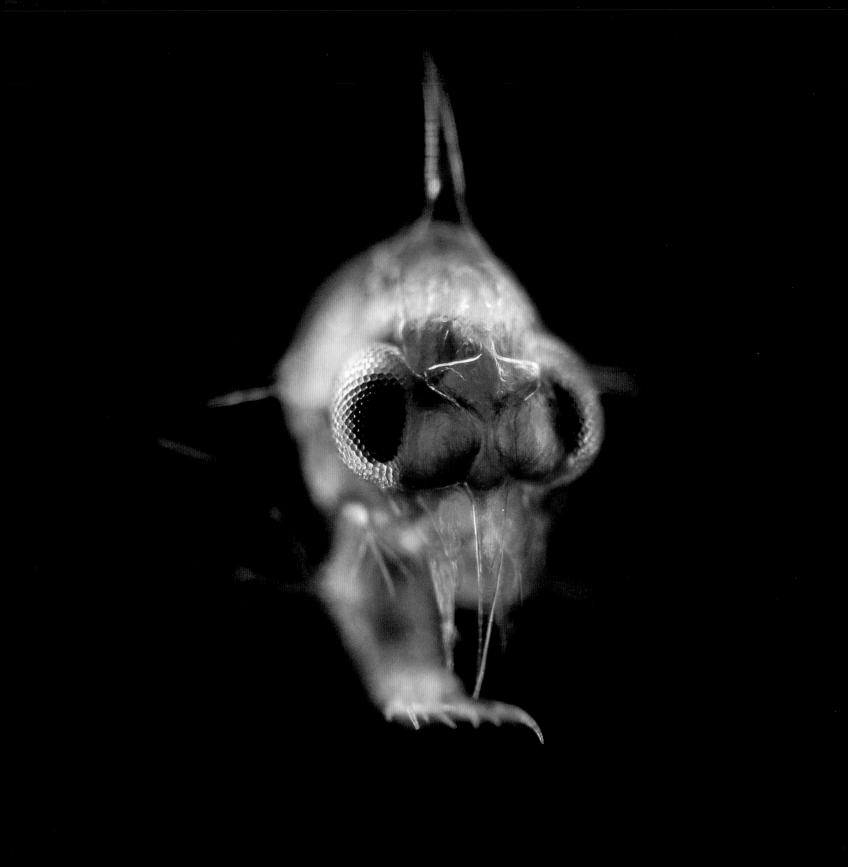

Barnacle larvae

Barnacles belong to the Crustacea, the same subphylum of the Arthropoda as the decapod crabs and shrimps. While most barnacles are gregarious species living in large groups firmly attached to rocks on the seashore or in the subtidal, they are a diverse group that also contains species living solely on the heads of whales or the shells of turtles. Some barnacles even bore into the shells of snails or are parasitic within crabs. It was a study of the diversity of barnacles and of how they are related to each other that helped Charles Darwin initially develop his theories on evolution. Most adult barnacles live upside down within a shell made of six calcite plates from where they extend their limbs into the water to catch food. Although most barnacles are hermaphrodites (both male and female) they usually have to mate with another barnacle. The fertilised eggs hatch into a nauplius larva that is brooded initially within the barnacle's shell until the first moult. The stage-2 nauplius is then released into the plankton where it feeds and moults a further five times before turning into the cyprid stage that is specialised to find a place to settle. The frontal horns on the carapace of this nauplius are a characteristic feature of the larvae of free-living barnacles.

The nauplius larva of the barnacle *Balanus perforatus*

MAGNIFICATION X 300

Biofouling

The final stage in the planktonic life of the barnacle is the bi-valved cyprid. The sole purpose of the cyprid, which doesn't feed, is to find somewhere suitable to settle. Barnacles have internal fertilisation and as the adults cannot move they must live within a penis-length of each other. Barnacles therefore use pheromones (chemical scents) to help the larvae settle near con-specifics (barnacles of the same species) so that they can mate. The two antennules protruding from the left of this cyprid enable it to sense pheromones given off by adult barnacles of the same species in order to determine where to settle. Because of their gregarious habit barnacles can be significant fouling organisms of underwater structures and of ships' hulls. Fouling of ships by barnacles and other marine organisms such as macroalgae (seaweed) affects the hydrodynamics of ships and can lead to an increase in fuel consumption of up to 40%. As a consequence, a large industry has emerged to develop ways to prevent the biofouling of submarine structures.

The cypris larva of the barnacle *Balanus perforatus*

MAGNIFICATION X 250

Fascinating facetotecta

This nauplius larva is still one of the mysteries of the plankton. Facetotectan larvae were first discovered around 1887 in the sea off the German coast and, although similar larvae have since been found in coastal waters around the world, still no one knows what they turn into as adults. Morphological and molecular evidence suggests that facetotectans are related to the barnacles as they share a similar pattern of development – they have a series of naupliar stages followed by a cyprid-like stage. Here, the similarities with free-living barnacles end, however. While free-living barnacle nauplii typically have a smooth carapace with prominent frontal horns, the facetotectan nauplius lacks horns and the carapace is covered in ridges giving a brick-like appearance. The facetotectan cyprid is also quite different in that the carapace comprises a single shell in contrast to the bi-valved shell of the free-living barnacles. Interestingly, some of the features of the facetotectan nauplius are similar to the larvae of parasitic barnacles that live inside other animals. The parasitic destiny of the facetotectan larva is also suggested by recent experiments on hormone-induced metamorphosis of the cyprid stage, which gives rise to an unsegmented, slug-like organism that is very similar to a life cycle stage seen in parasitic barnacles. A recent study of facetotectan larvae collected at just one site in coastal waters off Japan revealed more than 40 different species, which is an amazing biodiversity for this single obscure group and shows how little we still know about life in the sea.

Dorsal and ventral (showing the red eye spot) views of a facetotecan nauplius

MAGNIFICATION X 300

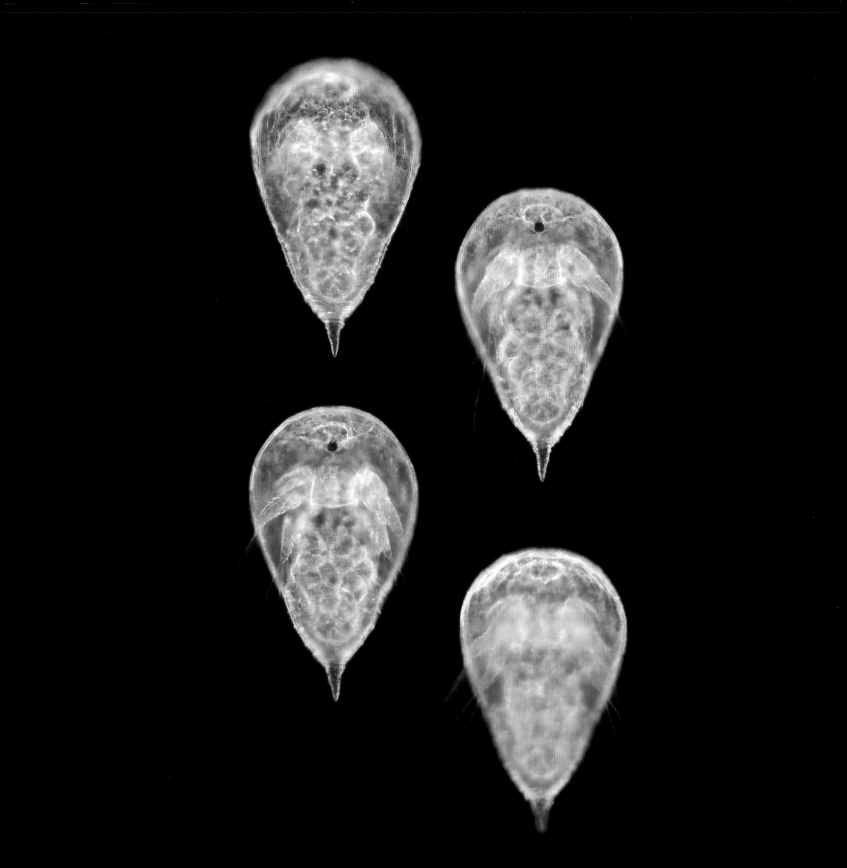

Young jellyfish

Many jellyfish, such as the scyphozoan moon jelly *Aurelia aurita*, have benthic and planktonic stages in their life cycle: an asexual benthic stage – the polyp – and a planktonic sexual stage – the medusa. The benthic polyp produces many ephyrae by an asexual budding process called strobilation (the ephyrae of *Aurelia aurita* shown here are only a few hours old). Each ephyra then grows into either a male or female planktonic medusa that eventually will release gametes (eggs or sperm) into the water. After reproduction, the fertilised eggs hatch into free-swimming planula larvae, which find somewhere to settle on the sea bed where they turn into benthic polyps to begin the life cycle again.

Ephyrae of the moon jellyfish *Aurelia aurita*

MAGNIFICATION X 20

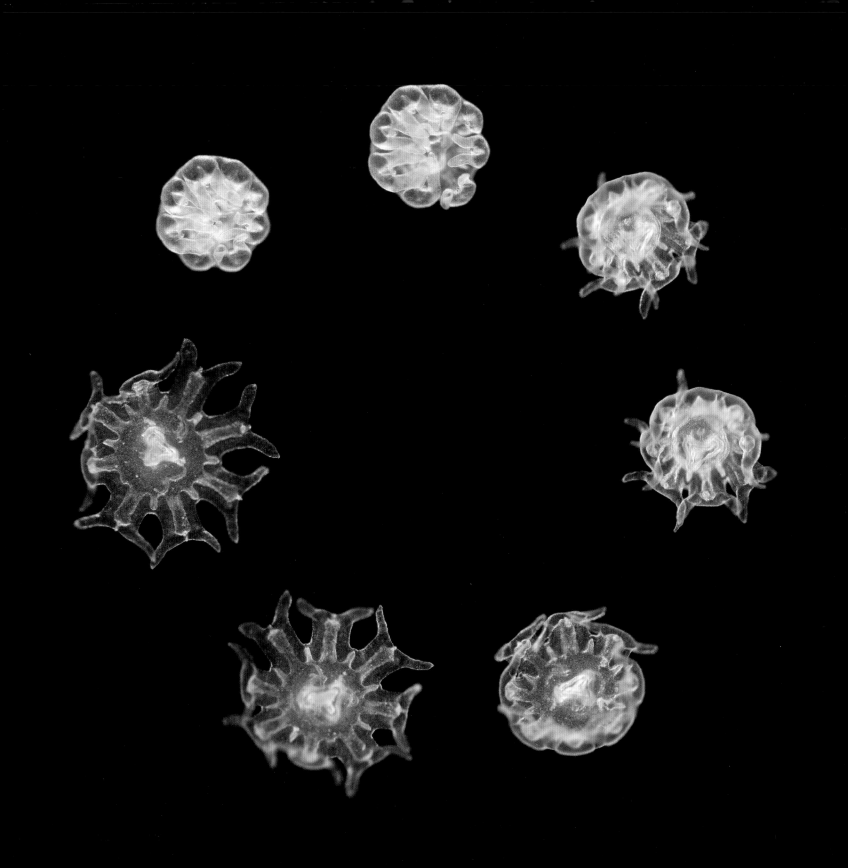

Moon jellyfish

Called the moon jelly, this 5 mm juvenile *Aurelia aurita* jellyfish will grow up to 40 cm across and, like all jellyfish, its body is made of a gelatinous tissue called the mesoglea that is greater than 95% water. *Aurelia* spp., occur in the plankton throughout the world's oceans where they feed on zooplankton and fish larvae, which they catch using harpoon-like stinging cells (nematocysts) located in the tentacles around the margin of the bell. In turn, moon jellies are the food of animals such as the leatherback turtle, *Dermochelys coriacea*, and the ocean sunfish, *Mola mola*. In recent years, the occurrence of large blooms of jellyfish seems to have increased in number throughout the oceans. Although the cause of this is still unclear, it is suggested that it may be due to a combination of climate change and the effects of overfishing.

A juvenile moon jellyfish *Aurelia aurita*

MAGNIFICATION X 30

Mini stinging predators

These tiny jellyfish are the sexual medusae of *Obelia* sp., a hydrozoan cnidarian. The Cnidaria also include the corals and the anemones (Anthozoa) and the 'true jellyfish' (Scyphozoa), such as *Aurelia aurita*. The name Cnidaria comes from the Greek word *cnidos*, which means 'stinging nettle'. As is the case with other Cnidaria, each of the stinging tentacles in these medusae is attached to a muscular band called the velum and they are used to catch small zooplankton as food. At the centre of the velum is the mouth (manubrium), which is surrounded by the gonads. These tiny medusae are themselves the prey of many zooplankton, such as the predatory copepod *Corycaeus anglicus*.

The medusae of *Obelia* sp.
MAGNIFICATION X 70

A medusa of *Obelia* sp.
MAGNIFICATION X 180

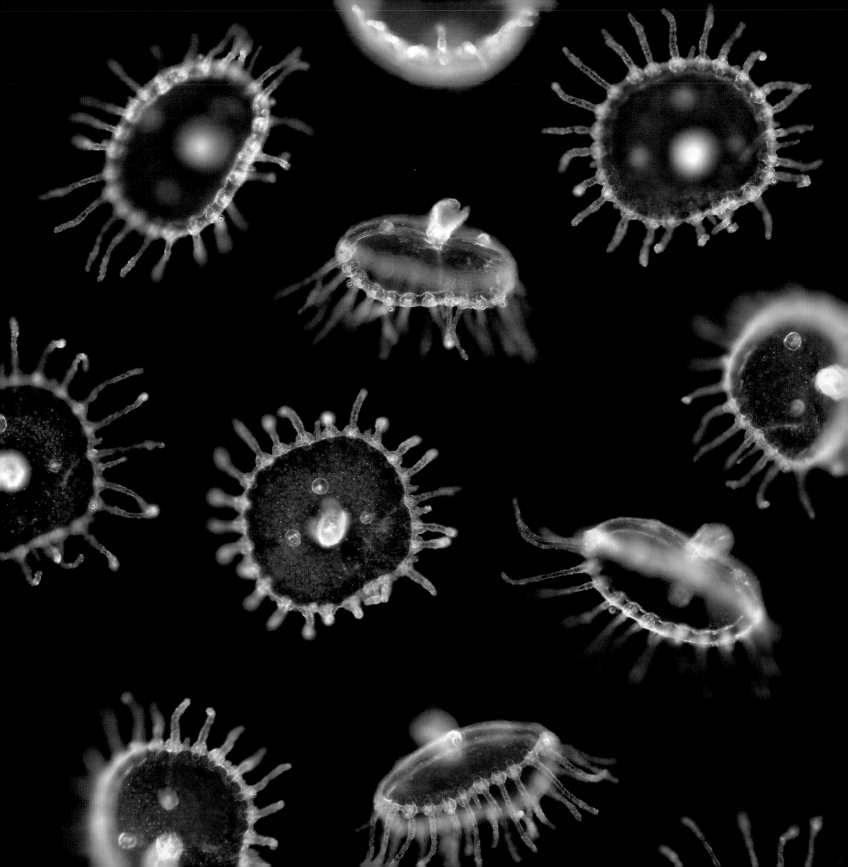

Jet propulsion

Like many Scyphozoa, Hydrozoa also have two life-history stages: a benthic hydroid stage that gives rise to many sexual planktonic medusae by an asexual budding process. The well-developed yellow gonads are very visible in this sexual stage of *Amphinema dinema*. Although they are largely at the whim of the ocean currents, the jellyfish like most zooplankton can swim to capture food or avoid predators. This hydrozoan jellyfish 'swims' by using contractions of the velum to expel a jet of water.

The medusae of *Amphinema dinema*

MAGNIFICATION X 30

One and the same

This jellyfish is the free-swimming, planktonic, sexual medusa of the hydrozoan *Lizzia blondina*, a benthic member of the Bougainvilliidae. As this jellyfish grows it passes through three stages that differ according to the number of tentacles they possess. The final stage is typified by four groups of three tentacles each interspersed by four single tentacles. Originally, the three different stages were considered to be separate species called, *Dysmorphosa minima*, *Lizzia claparedei* and *Lizzia blondina*. Now, it is appreciated they are all the same animal. *Lizzia blondina* can reach large numbers in the plankton by budding off new medusae from the stomach wall and in this way they may become the dominant medusae in coastal waters where they occur.

The medusa of *Lizzia blondina*

MAGNIFICATION X 40

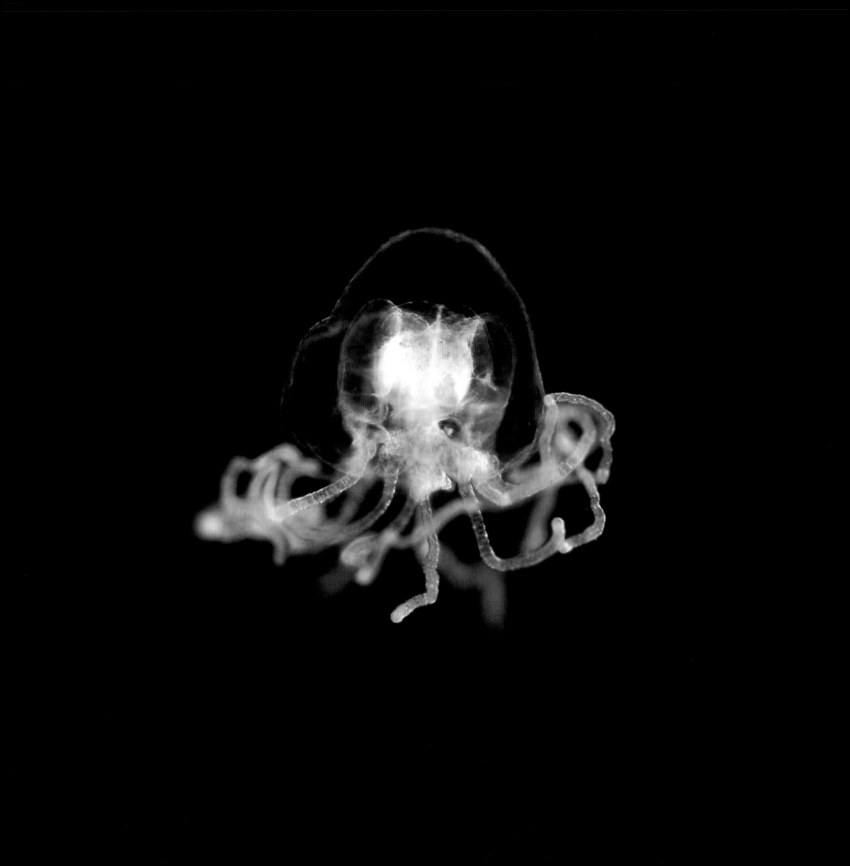

Sea urchins

This eight-armed creature is a late-stage echinopluteus larva of a regular (radially symmetrical) sea urchin. Sea urchins are echinoderms, a phylum that also includes the starfish, sea cucumbers and crinoids (feather stars). Echinoderm larvae are among the smallest of the merozooplankton. As the larva grows, the number of arms it possesses gradually increases. Each arm has an internal skeletal rod of calcite (calcium carbonate) covered by an epithelium (skin) that bears beating cilia that the larva uses for locomotion and feeding. In many areas of the world the adults of some species of regular sea urchin, such as *Paracentrotus lividus* and *Evechinus chloroticus*, are harvested for their roe (eggs), which are considered a culinary delicacy.

An echinopluteus larva of the sea urchin *Echinus esculentus*

MAGNIFICATION X 120

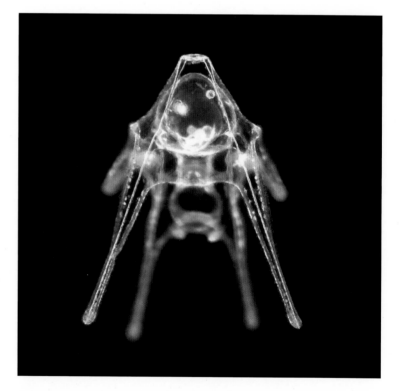

The echinopluteus larva of *Psammechinus miliaris*

MAGNIFICATION X 240

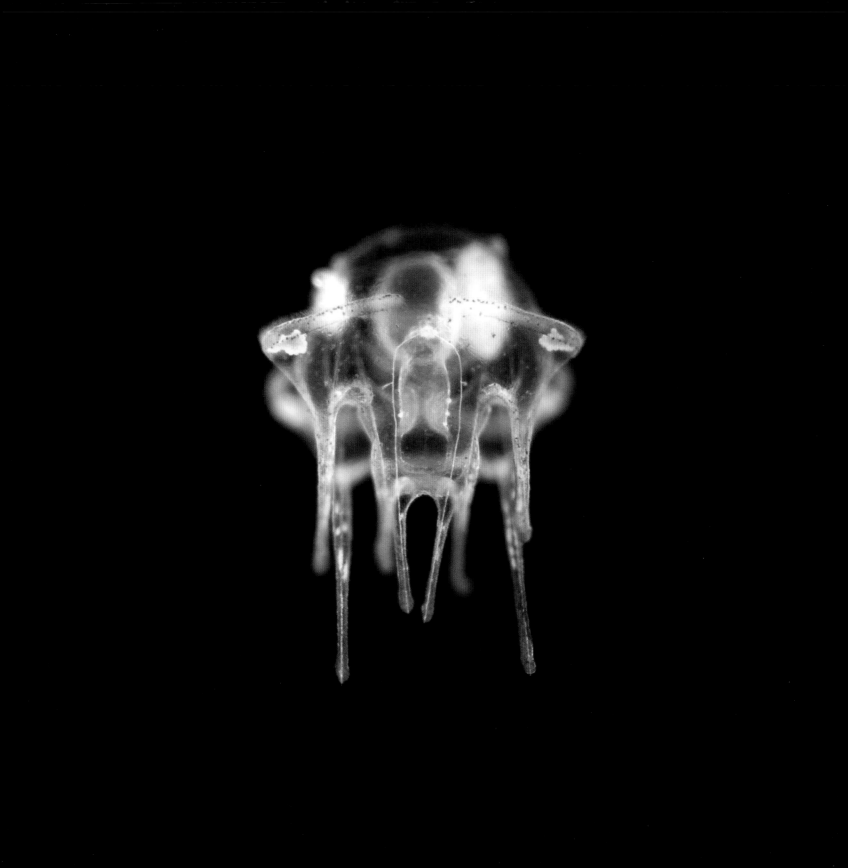

Heart urchins

This is the echinopluteus larva of the irregular urchin *Echinocardium cordatum*. Irregular urchins differ to the radially symmetrical regular urchins in that, as adults, they show secondary bilateral symmetry. In contrast to regular urchins such as the common sea urchin *Echinus esculentus*, where the mouth and anus are opposite each other, the anus is shifted posteriorly in irregular urchins. Irregular urchins generally burrow into mud or soft sediments where they feed on dead and decaying organic matter and play an important role in bioturbation (the turning over of sediments) and nutrient recycling.

The echinopluteus larva of the heart urchin *Echinocardium cordatum*

MAGNIFICATION X 200

Brittle stars

Like the echinopluteus sea urchin larva, this ophiopluteus larva of a brittle star has projecting arms supported by skeletal rods around which loops a band of beating cilia. The larva swims with the arms pointing forward so that the hair-like cilia create a current that sweeps food particles towards the central mouth. At settlement and metamorphosis into the young brittle star on the sea bed the larval tissues are absorbed. Brittle stars live in dense aggregations on the sea bed where they play an important role in nutrient recycling by feeding on decaying phytoplankton and other organic matter that sinks from the sea surface to the sea floor.

The long-armed ophiopluteus larva of *Ophiothrix fragilis*

MAGNIFICATION X 500

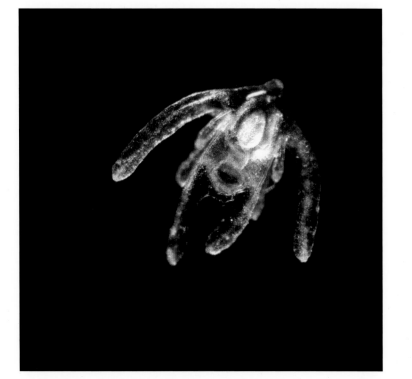

A short-armed ophiopluteus larva

MAGNIFICATION X 300

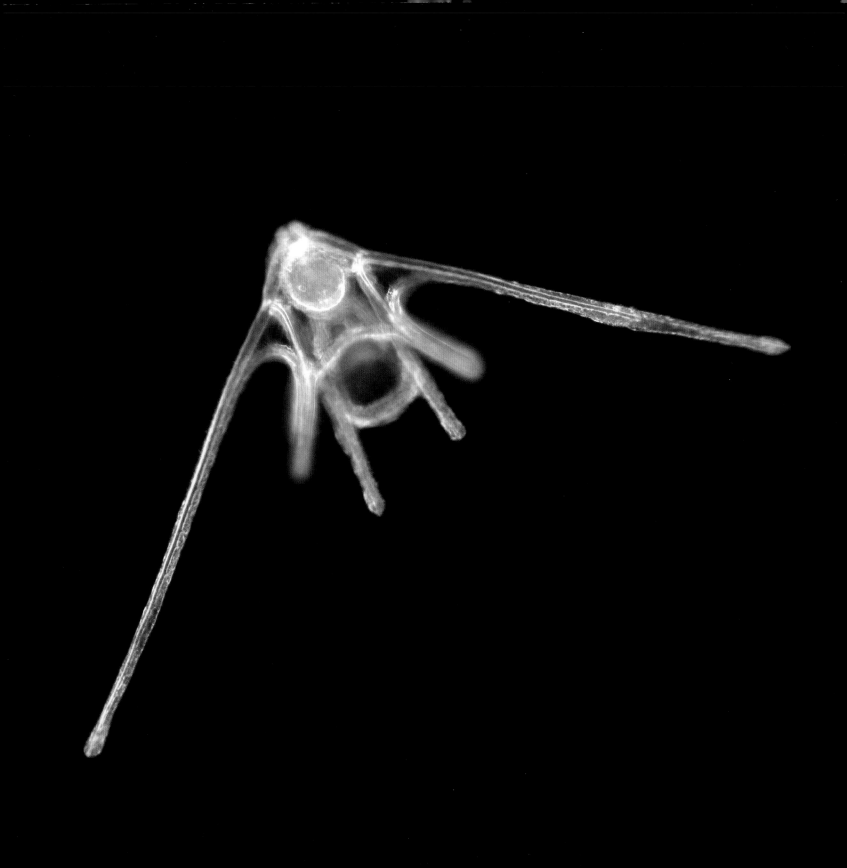

Sea bed hunters

This late-stage brachiolaria larva of an asteroid echinoderm will become a starfish on the sea bed. The boundaries between each of the five developing arms of the juvenile starfish are clearly marked by dimples in this larva. Many adult starfish are important hunters on the sea bed where they feed on mussels, crabs and even other species of starfish. Starfish can occur in high densities on the sea bed and as a swarm travels across the surface they leave virtually nothing alive behind them. Most adult starfish have an unusual method of feeding that involves turning their stomach completely inside out. After capturing their prey, such as a mussel, the starfish will first prise open the shell using its powerful tube feet situated in rows along each arm. Next, it will evert its stomach through the mouth to envelope the prey, which is then digested externally. After the stomach lining has absorbed the liquefied prey it is drawn back through the mouth and into the starfish's body.

A late-stage brachiolaria larva of a starfish

MAGNIFICATION X 180

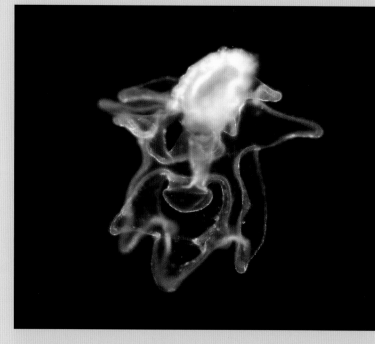

Early-stage brachiolaria larva

MAGNIFICATION X 250

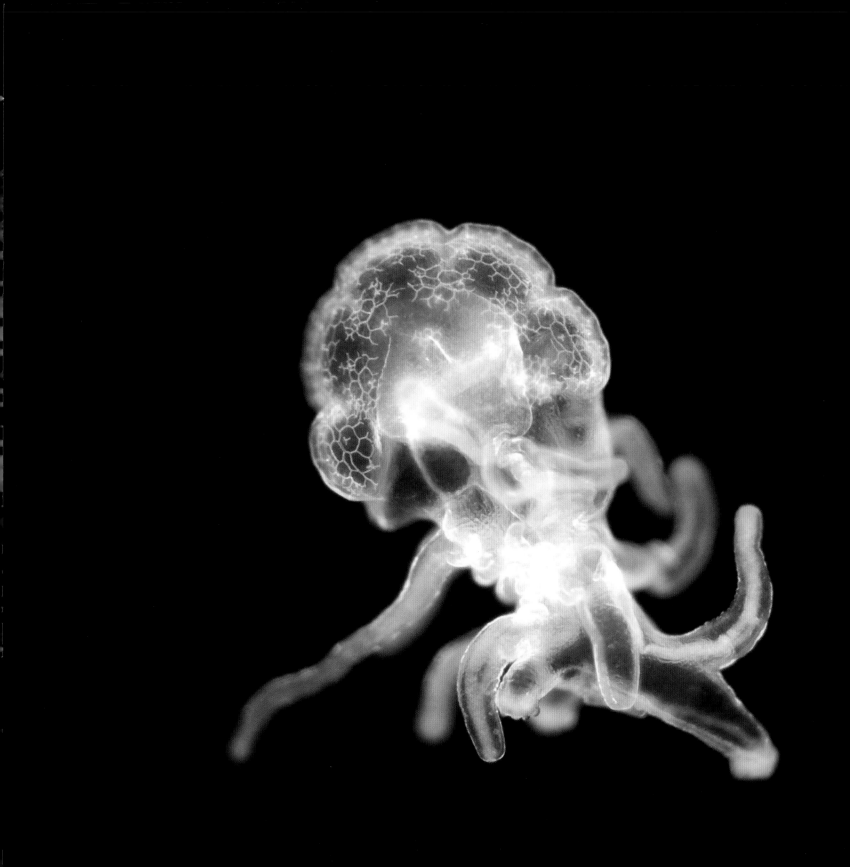

Born to be different

Luidia sarsi has an unusual larval development among echinoderms. In most starfish and sea urchins the larval tissues are absorbed when the larva metamorphoses into the juvenile that settles to the sea floor. In *L. sarsi*, however, the developing yellow-orange juvenile detaches from the larval body and sinks to the sea bed. The rest of the larval body may then continue to swim in the plankton for more than a month before it runs out of energy and dies. Like many starfish, adult *L. sarsi* have the remarkable ability to regenerate lost arms, or even grow a whole new individual from a single arm, provided a part of the central body remains attached.

The larva of *Luidia sarsi*

MAGNIFICATION X 80

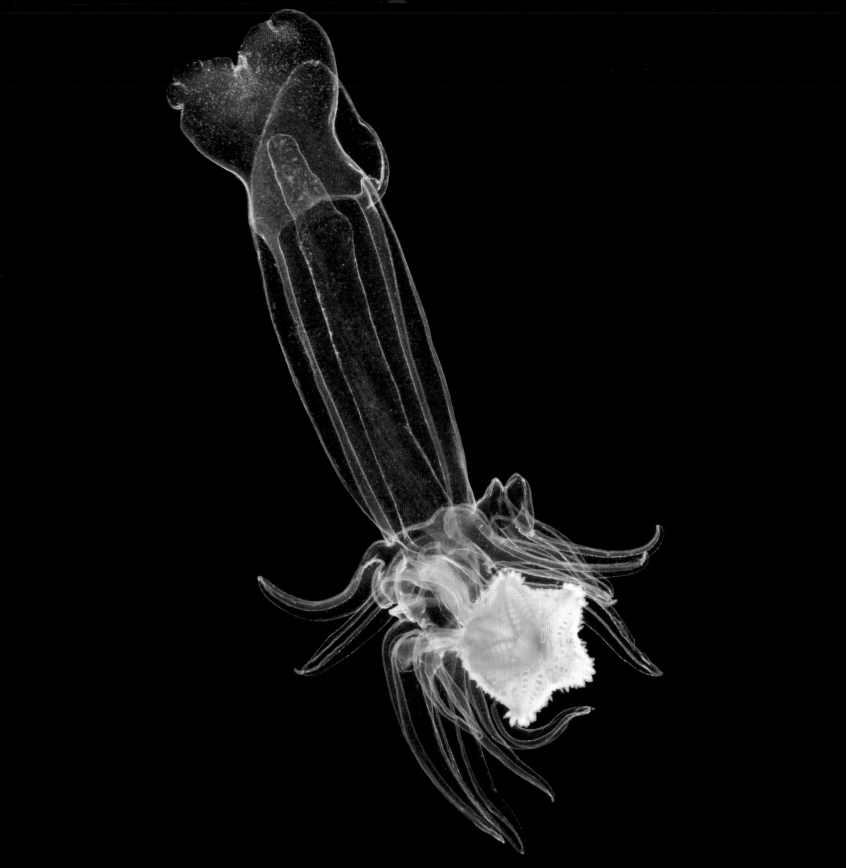

Settling down

Just like its close relative, *Luidia sarsi*, the juvenile of *L. ciliaris* detaches from the larval body and sinks gradually to the sea bed. This sinking juvenile *L. ciliaris* has seven arms in contrast to the five arms of *L. sarsi*. The adult can grow to 60 cm in diameter and is red-orange in colour with a row of white teeth-like spines along the side of each arm. These starfish are active, fast-moving carnivores on the sea bed preying on other echinoderms, especially brittle stars.

A juvenile *Luidia ciliaris*

MAGNIFICATION X 130

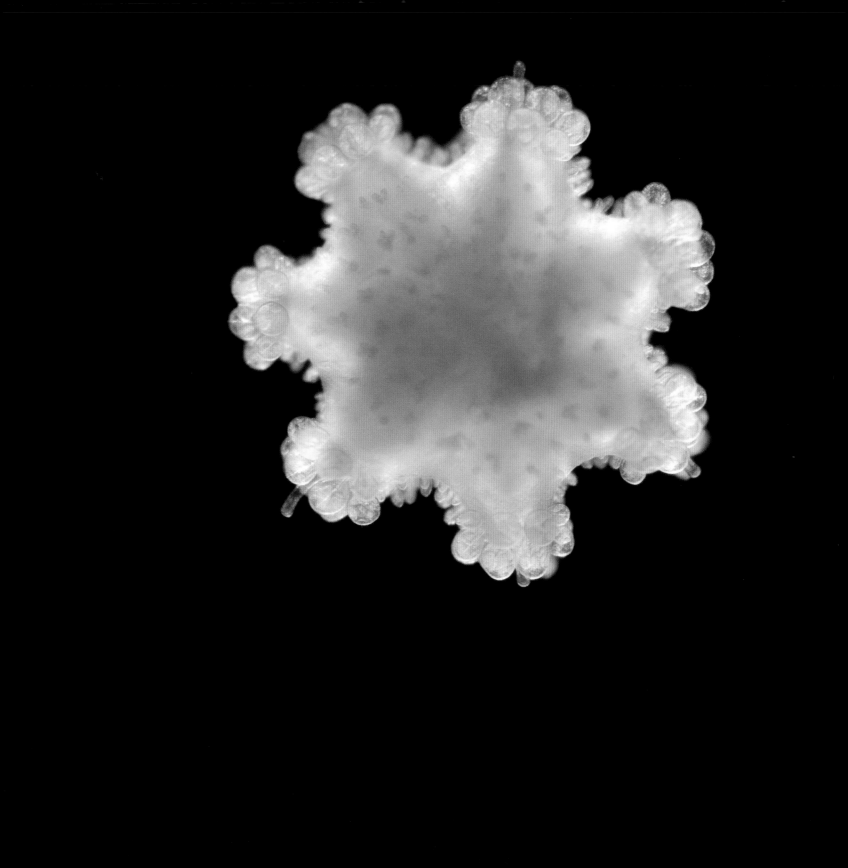

Sea cucumber auricularia larvae

This is the auricularia larva of a holothuroid or 'sea cucumber', so-called because of the surface texture and elongate shape of the adult. Sea cucumbers are also echinoderms, a name that comes from the Greek word meaning 'spiny skin' since all adult echinoderms have an internal scaffold-like skeleton made of tiny calcium carbonate plates. Some of the developing skeletal ossicles of the juvenile sea cucumber can be seen near the lower left margin of this larva. The convoluted folds of this larva are fringed with beating cilia that are used for locomotion and to generate a current that draws food particles towards the mouth in a similar fashion to the echinopluteus and ophiopluteus larvae of sea urchins and brittle stars. With the exception of a few pelagic species, most adult holothuroids live on the sea bed where they are detritivores feeding on decaying plankton that has sunk to the sea floor.

The larva of a sea cucumber

MAGNIFICATION X 340

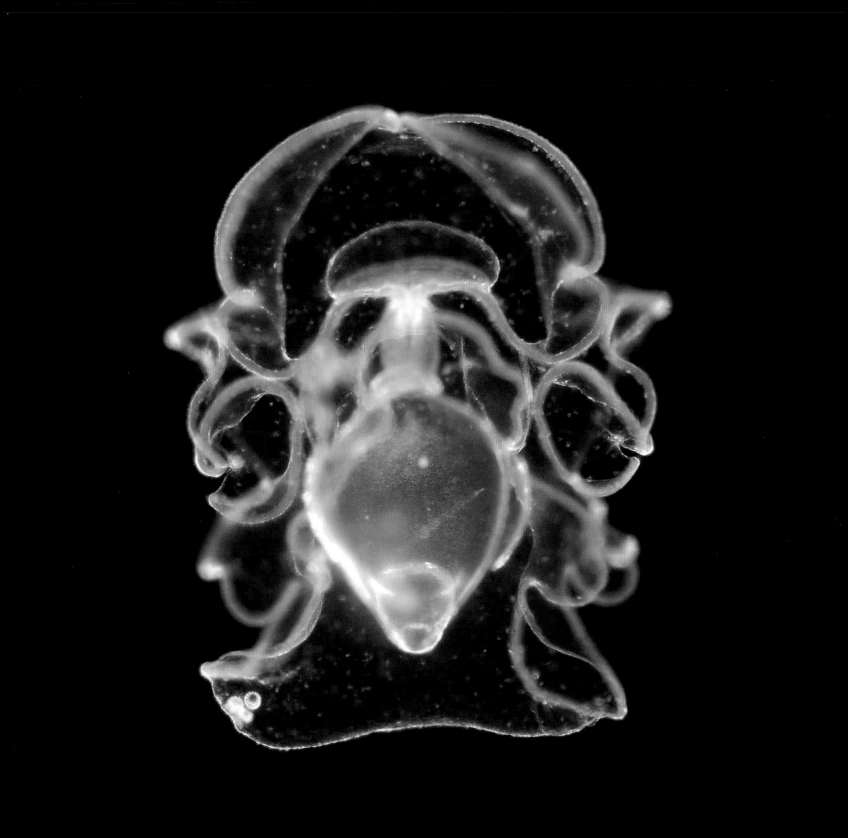

Sea cucumber pentacula larvae

Sea cucumbers show both direct and indirect development. Species with direct development produce either a benthic or planktonic non-feeding vitellaria larva that develops into an adult individual. In contrast, species with indirect development possess three distinct larval stages, of which only the first, the auricularia, is truly planktonic. The auricularia stage, after feeding and growing in the plankton, metamorphoses into a barrel-shaped doliolaria larva that soon becomes the settlement-stage pentacula that sinks slowly from the plankton to the sea floor. The pentacula has five motile tentacles that it uses to 'walk' around on the sea bed.

The pentacula larva of a sea cucumber

MAGNIFICATION X 120

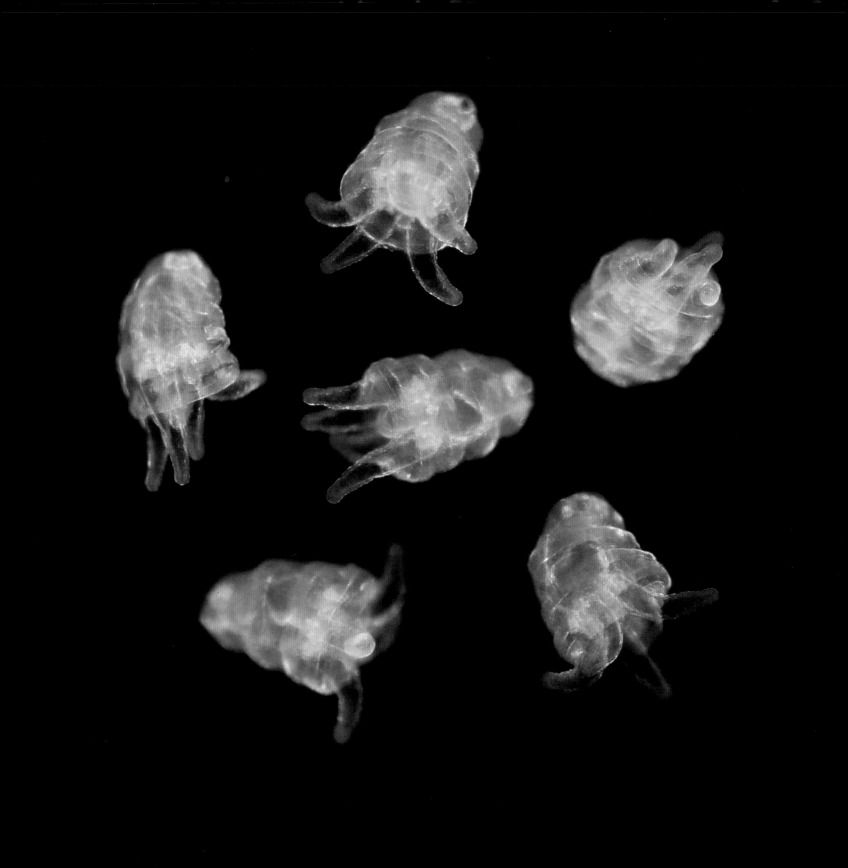

Acorn worms

This is the tornaria larva of an enteropneust or 'acorn worm' that, as an adult, will live in a mucus-lined burrow in the sea bed. Although they are called worms because of their worm-like appearance, acorn worms are hemichordates, a phylum that is exclusively marine. The Hemichordata is considered to be the sister phylum to the Echinodermata and this young tornaria larva, with its pre-oral ciliary band that loops around the body, can be seen to show some similarities with the larvae of echinoderms, such as the larvae of sea cucumbers and starfish. Indeed, for many years the tornaria was thought to be a larval echinoderm. Just like the larvae of echinoderms, the band of cilia generates a feeding current to convey food to the mouth that in the tornaria is located in a groove on the side of the body. To help this tornaria larva swim it also has a telotroch – a second band of large perianal compound cilia that is used solely for propulsion.

Earlier and later-stage tornaria larva of an acorn worm

MAGNIFICATION X 120

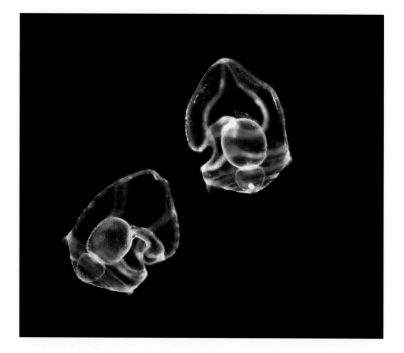

Young tornaria larvae

MAGNIFICATION X 80

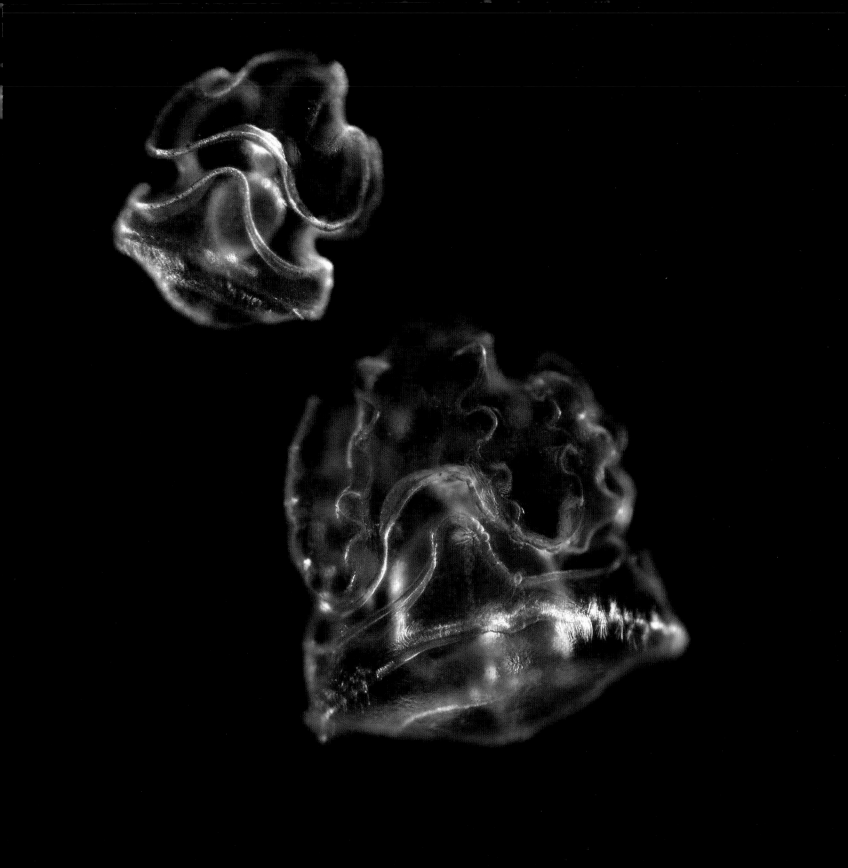

Horseshoe worms

This is the actinotrocha larva of a phoronid or 'horseshoe worm', also
known as the phoronid larva. Adult horseshoe worms are benthic animals
that live in tubes and are so-called because of the horseshoe shape of their
feeding fan, which they orientate into the water current to catch food
particles. When this larva was first described by Müller in 1845 it was
believed to be an adult animal in its own right and it was given the name
Actinotrocha branchiata. In 1867, Kowalevsky realised that it was actually
the larva of a phoronid worm. The larva retains two different names to this
day. There are only two genera within the Phoronida and there are only 20
known species worlwide.

The actinotrocha or phoronid larva of *Phoronis muelleri*

MAGNIFICATION X 200

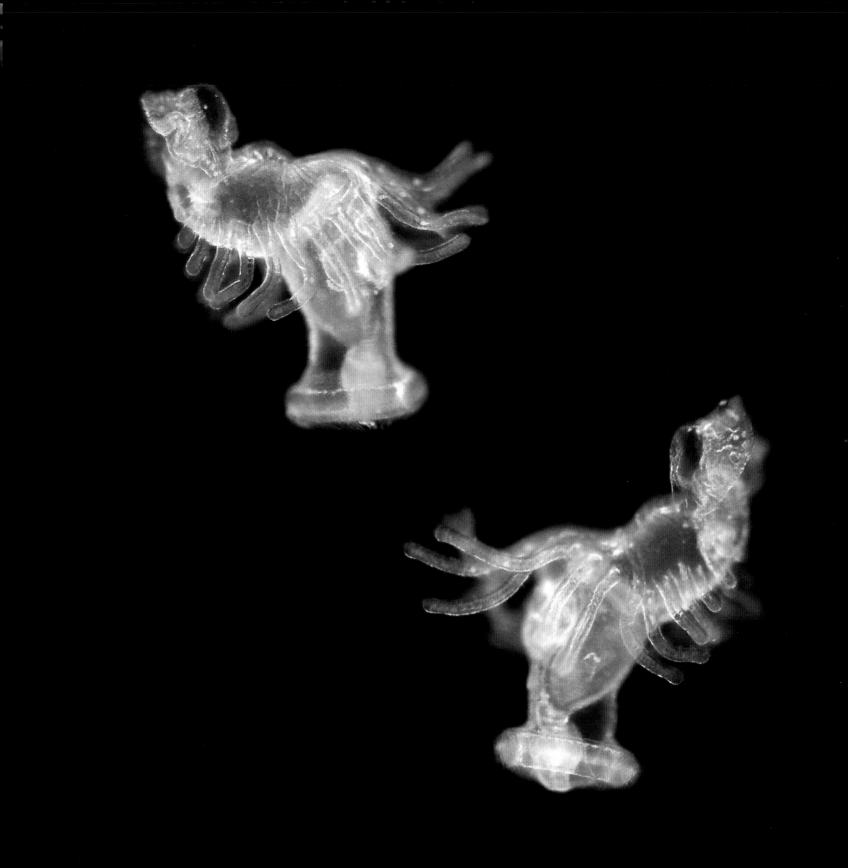

Moss animals

The cyphonautes is the planktonic larva of a bryozoan (phylum Ectoprocta). Adult bryozoans form large colonies by asexually budding new individuals (zooids) and they can be found encrusting rocks, the shells of other marine animals such as bivalves, and the fronds of seaweeds. Adult bryozoans are known colloquially as 'moss animals' or 'sea mats'. Most bryozoans are hermaphrodites and brood the fertilised eggs before releasing juvenile cyphonautes larvae into the water column. Like the adult, the larva feeds on particles floating in the water. The band of motile cilia around the corona of the pyramid-like shell of these larvae is used for both swimming and to draw in water. Inside the shell, within the mantle cavity, stiff and stationary cilia act like a comb to sieve small food particles from the current of water flowing through.

The cyphonautes larva

MAGNIFICATION X 120

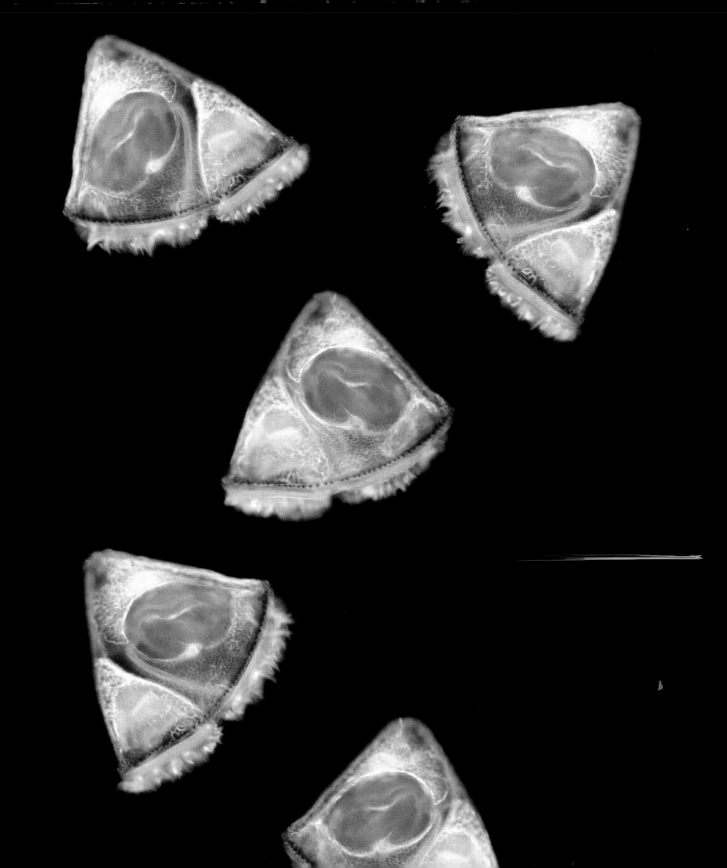

Ribbon worms

This is the pilidium larva of a ribbon worm belonging to the phylum Nemertea. Unlike true worms, which belong to the Annelida, ribbon worms lack any segmentation. Most nemerteans are predatory marine animals living in sediments on the sea bed. While the egg of many nemertean species hatches directly into the juvenile worm, some nemerteans have a planktonic larval stage called the pilidium. Looking like a small helmet, the pilidium larva has a tuft of long cilia at its top and pair of ciliated lobes hanging below that generate a feeding current, and help it to swim.

The pilidium larva of a ribbon worm

MAGNIFICATION X 220

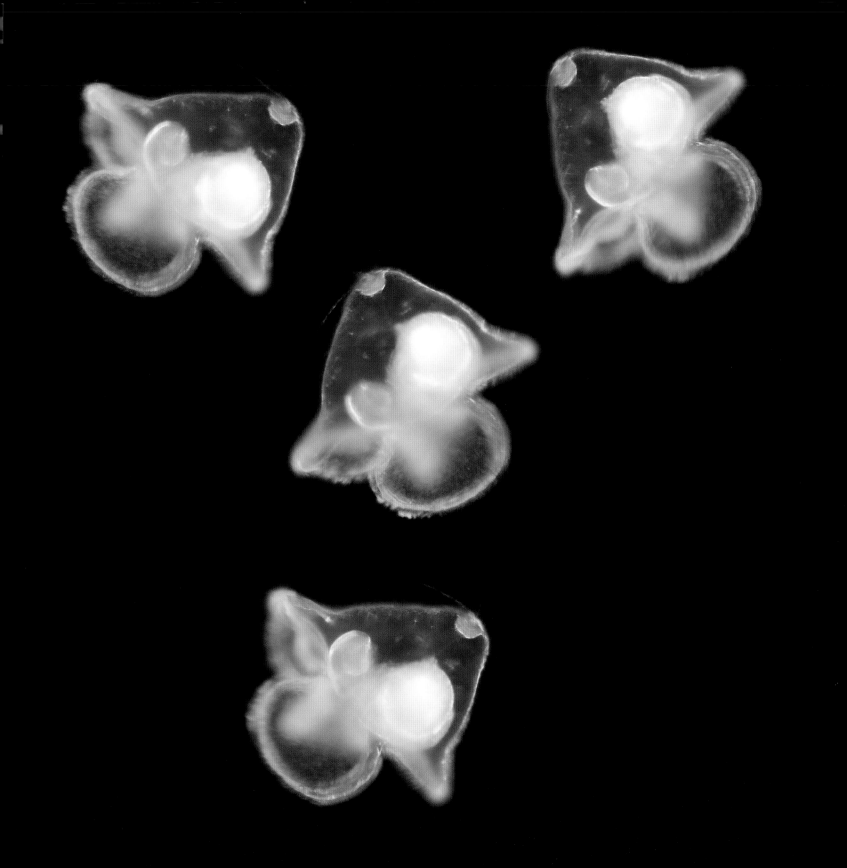

Worm metatrochophores

The metatrochophore larva of a polychaete worm has three bands of beating cilia: two that are closely positioned – the prototroch (a pre-oral band of cilia) and the metatroch that together flank a groove just below the head – and a terminal telotroch. The beating of these cilia generates a feeding current and propels the larva in a corkscrew-like fashion through the water. The two eyespots on the head and the beginnings of the segmentation that will characterise the adult worm can just be seen in these larvae. As this metatrochophore larva grows older the ciliary bands will be lost, the segmentation will become more developed and bristle-like chaetae will grow. The larval stage prior to the metatrochophore is known as the trochophore.

An annelid metatrochophore larva

MAGNIFICATION X 160

The mitraria larva of an owenid polychaete

MAGNIFICATION X 170

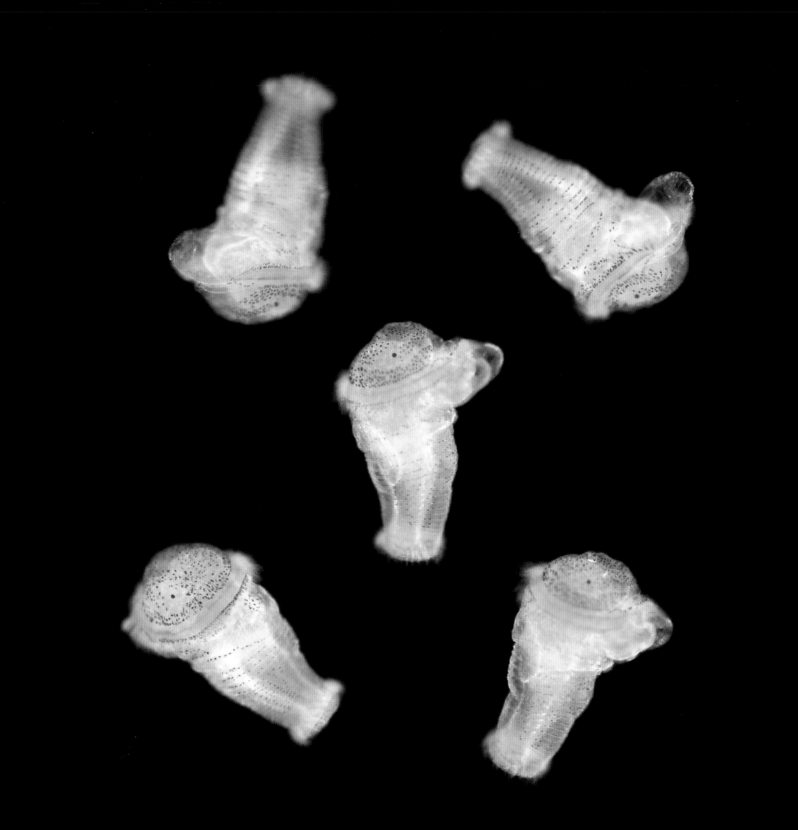

Bristling with bristles

Polychaetes are sometimes called bristle worms because of the bundles of chaetae they possess. The adult worms mostly live in muddy and silty sediments where they are tube builders. These worms provide a food source on intertidal mudflats for many fish and, during low tide, for wading birds. The chaetae of this free-swimming spionid polychaete larva can be erected rapidly and are thought to act as both a defence mechanism against predators, such as crab and fish larvae, and to help them stay afloat in the plankton.

The larva of a spionid polychaete

MAGNIFICATION X 150

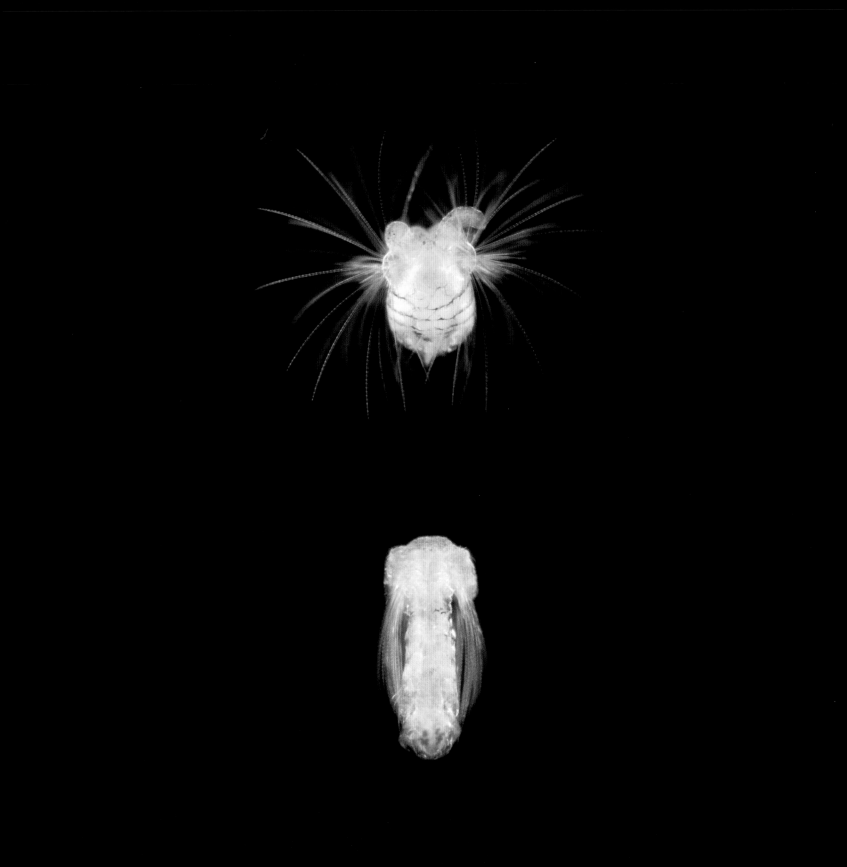

Spionid polychaete

There are lots of different species of spionid polychaete and, like in this *Prionospio malmgreni*, each bundle of chaetae projects from paired parapodia that arise from each side of a body segment. Parapodia are paddle-like, fleshy lobes containing lots of blood vessels and they are used for both respiration and locomotion. In spionid worms, and most other polychaetes, the bristles are made of chitin. However, in the tropical polychaetes known as fireworms, the chaetae are made of calcium carbonate and are hollow. When a fireworm is attacked the brittle chaetae break releasing a neurotoxin that acts as a powerful deterrent against predation.

The spionid polychaete *Prionospio malmgreni*

MAGNIFICATION X 100

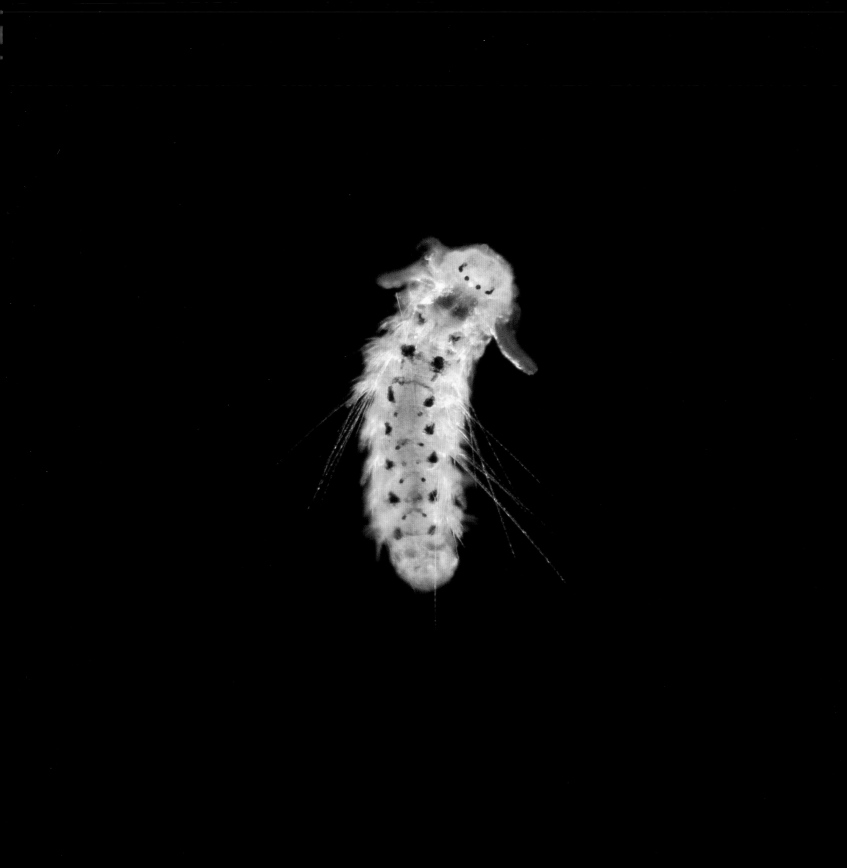

Parchment worms

This is the larva of the polychaete parchment worm, *Chaetopterus variopedatus*, so called because of the papery tube the adult worm builds in the sand. The larva has a wide mouth and the body is segmented like all polychaete larvae. Clearly visible around the middle of this larva are the two bands of beating cilia that it uses to swim. Also visible on the head are the lateral eyespots. Within the larva can be seen several brown faecal pellets. Plankton faecal pellets play an important role in the biological carbon pump because they are dense and therefore sink rapidly to the sea bed taking the carbon they contain with them. The adult parchment worm is arguably one of the most spectacular of all polychaetes. The U-shaped papery tube the adult builds is about 60 cm long and up to 4 cm in diameter. Most of the tube is buried in the sand with its openings projecting above the surface like two chimneys. The adult worm has many distinct segments bearing limb-like appendages called notopodia, some of which fit the tube precisely and act like a piston to draw in water and food particles as the animal moves backwards and forwards. On the 12th segment behind the head are two large, wing-like structures that secrete a mesh-like sheet of mucus that is shaped into a bag. As the water flowing though the tube passes through this mucus sieve any food particles are trapped. As soon as the mucus bag is full with food it is rolled up into a ball and moved forwards to the mouth to be eaten.

The larva of *Chaetopterus variopedatus*

MAGNIFICATION X 140

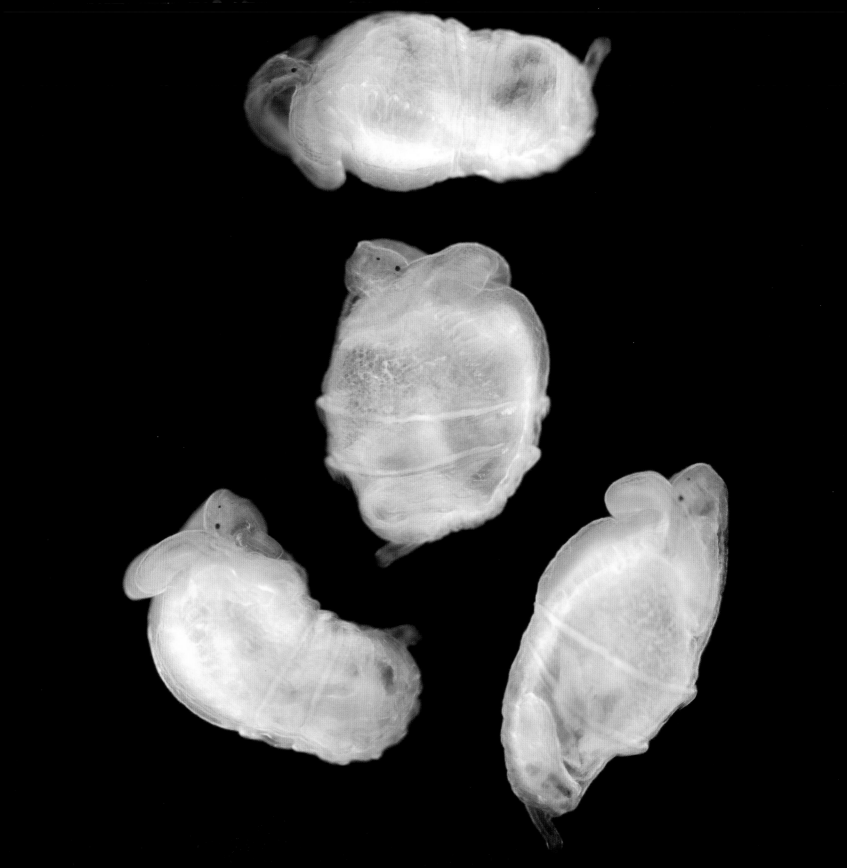

Life in a tube

To protect its soft body this late-stage worm larva has a tube to live within. Many adult polychaetes also build tubes to live within, which may be made from secretions or from imbricated (brick-like) said grains held together by mucous. Some species live alone in the sand, for example *Amphitrite johnstoni*, while others, like *Sabellaria alveolata*, construct large reef-like colonies that themselves become a habitat for other animals. The tentacle-like protrusions from the head of this larva, which is probably a terebellid Ampharetidae, are its gills and are not involved in feeding, the true feeding tentacles of the adult will develop from within the larva's mouth.

The planktonic larva of a tube-building worm

MAGNIFICATION X 200

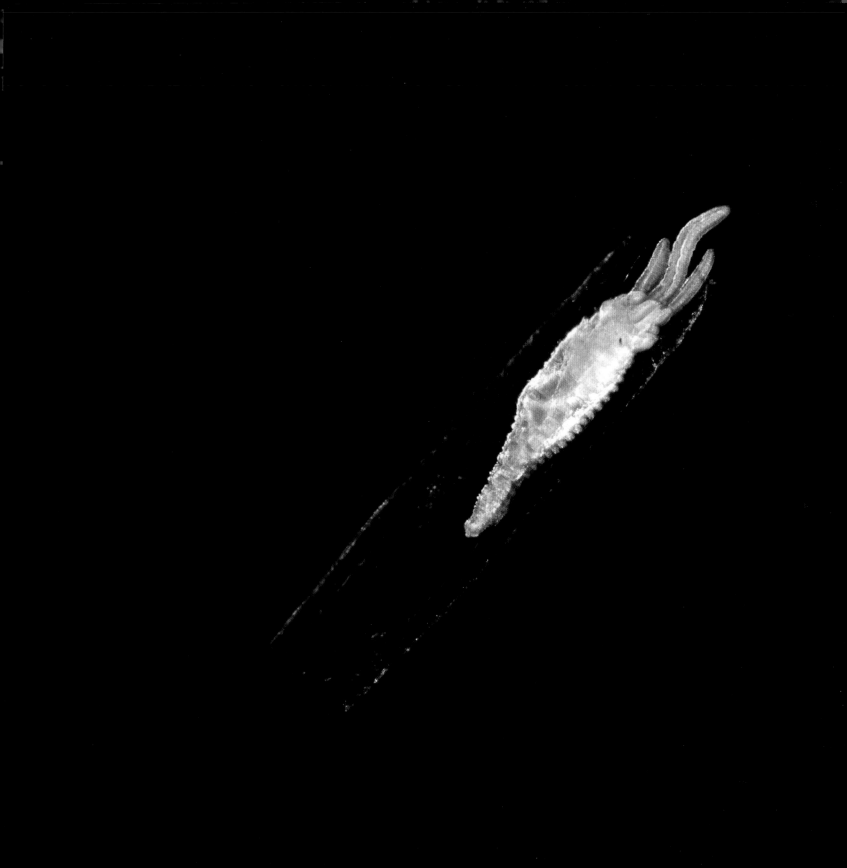

Spaghetti worms

Most adult Terebellidae live on the sea floor in delicate membraneous tubes buried in the mud or sand and they are sometimes called fan or spaghetti worms because of their spaghetti-like feeding tentacles. The adult worms are deposit feeders that collect food particles by extending their tentacles over the sediment surface where their mucus coated surfaces pick up decaying plankton that has sunk to the sea bed. In this way, these worms, like many other detritivores on the sea bed, play an important role in nutrient cycling and, through the movement in their burrows, also serve to aerate the sediment surface.

The larva of a terebellid worm

MAGNIFICATION X 120

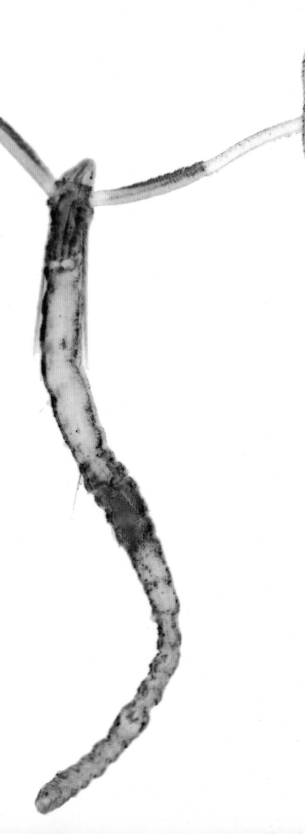

Magelonids

The long delicate palps of this juvenile polychaete are almost as long as its body. The adult *Magelona* sp. lives in fine, muddy sand with its long palps, which have a sticky surface, protruding into the water where they collect food particles. Because the palps wave around in the water column above the burrow, they are often bitten off and eaten by fish. Fortunately, for the adult worm, the palps have the ability to re-grow.

A juvenile magelonid polychaete

MAGNIFICATION X 120

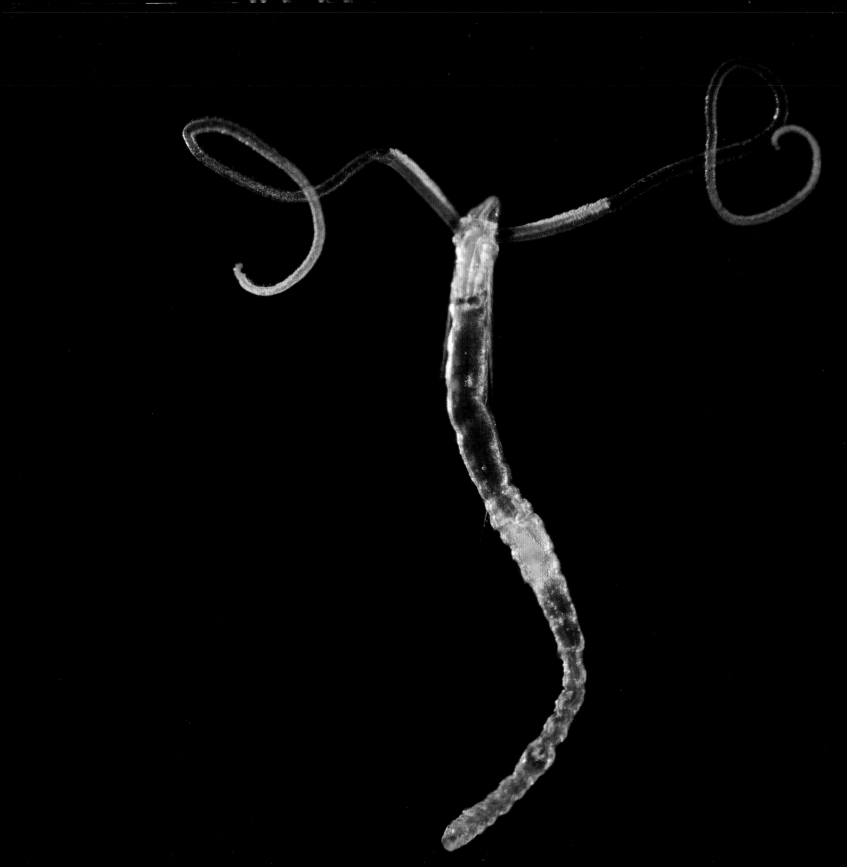

Snail veliger larvae

The gastropod veliger has a foot, a shell to hide in and an operculum to close the shell's opening, just like the adult snail it will become. To feed and stay afloat this veliger has two large velar lobes each fringed with beating cilia. One of the first scientists to work at the Marine Biological Association of the UK when it opened on Plymouth Hoe in 1888 was Walter Garstang. Garstang studied the evolution of marine larval adaptations and one of his hobbies was describing larval evolution in verse. His most famous poem is called *The Ballad of the Veliger*, or how the gastropod got its twist. In rhyming verse, Walter Garstang describes a unique property of gastropod growth known as torsion, a 180 degree twist of the body during larval development, which brings the animal's posterior to lie just above its head.

The veliger larva of *Nassarius (Hinia) incrassatus*

MAGNIFICATION X 100

The veliger larva of *Rissoa* sp.
MAGNIFICATION X 180

The Ballad of the Veliger
Or how the gastropod got its twist

The veliger's a lively tar, the liveliest afloat,
A whirling wheel on either side propels his little boat;
But when the danger signal warns his bustling submarine,
He stops the engine, shuts the port, and drops below unseen.

He's witnessed several changes in pelagic motor-craft;
The first he sailed was just a tub, with a tiny cabin aft.
An Archi-mollusk fashioned it, according to his kind,
He'd always stowed his gills and things in a mantle-sac behind.

Young Archi-mollusks went to sea with nothing but a velum –
A sort of autocycling hoop, instead of pram – to wheel 'em;
And, spinning round, they one by one acquired parental features,
A shell above, a foot below – the queerest little creatures.

But when by chance they brushed against their neighbours in the briny,
Coelenterates with stinging threads and Arthropods so spiny,
By one weak spot betrayed, alas, they fell an easy prey –
Their soft preoral lobes in front could not be tucked away!

Their feet, you see, amidships, next the cuddy-hole abaft,
Drew in at once, and left their heads exposed to every shaft.
So Archi-mollusks dwindled, and the race was sinking fast,
When by the merest accident salvation came at last.

A fleet of fry turned out one day, eventful in the sequel,
Whose left and right retractors on the two sides were unequal:
Their starboard halliards fixed astern alone supplied the head,
While those set aport were spread abeam and served the back instead.

Predaceous foes, still drifting by in numbers unabated,
Were baffled now by tactics which their dining plans frustrated.
Their prey upon alarm collapsed, but promptly turned about,
With the tender morsal safe within and the horny foot without!

This manoeuvre (vide Lamarck) speeded up with repetition,
Until the parts affected gained a rhythmical condition,
And torsion, needing now no more a stimulating stab,
Will take its predetermined course in a watchglass in the lab.

In this way, then, the Veliger, triumphantly askew,
Acquired his cabin for'ard, holding all his sailing crew –
A Trochosphere in armour cased, with a foot to work the hatch,
And double screws to drive a head with smartness and despatch.

But when the first new Veligers came home again to shore,
And settled down as Gastropods with mantle-sac afore,
The Archi-mollusk sought a cleft his shame and grief to hide,
Crunched horribly his horny teeth, gave up the ghost, and died.

W. Garstang 1928

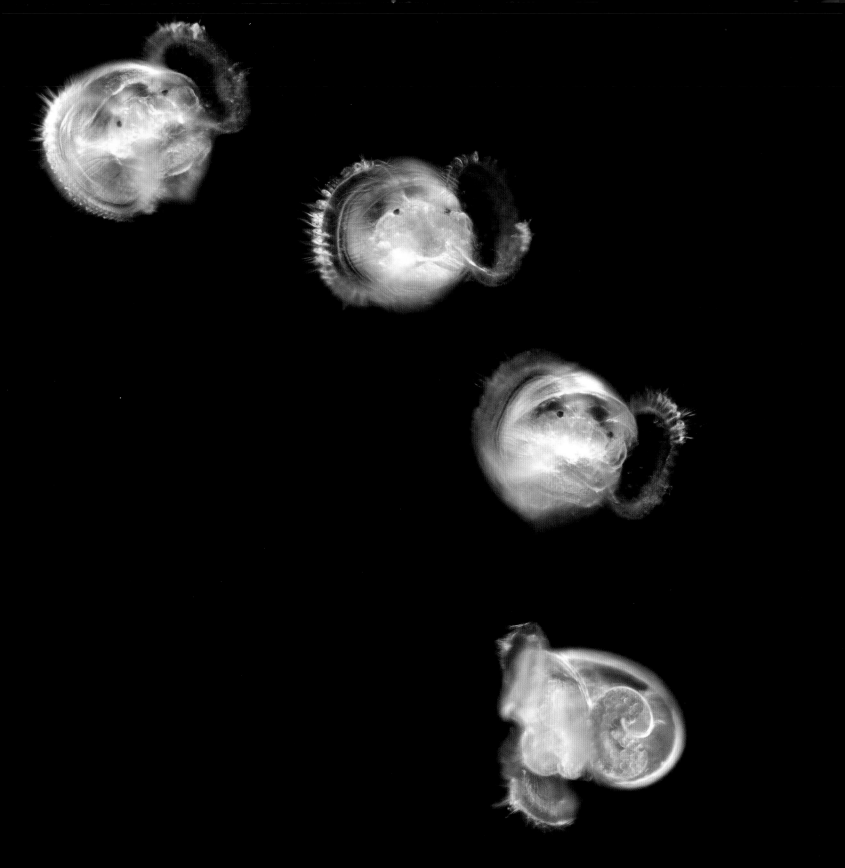

Juvenile bivalves

These tiny bivalves are no longer larvae and will soon be heavy enough to sink to the sea floor and grow into adults. Bivalves play an important role on the sea bed where some are deposit feeders, like clams, hoovering up dead and decaying plankton from the surface of the sediment, while other species, like mussels, filter feed on phytoplankton living in the water column. Several species of bivalves, such as the mussel *Mytilus edulis* and the oyster *Ostrea edulis,* are farmed commercially for our food. On the sea bed these bivalves are an important food source for many other benthic species such as crabs and flatfish.

Juvenile bivalves

MAGNIFICATION X 100

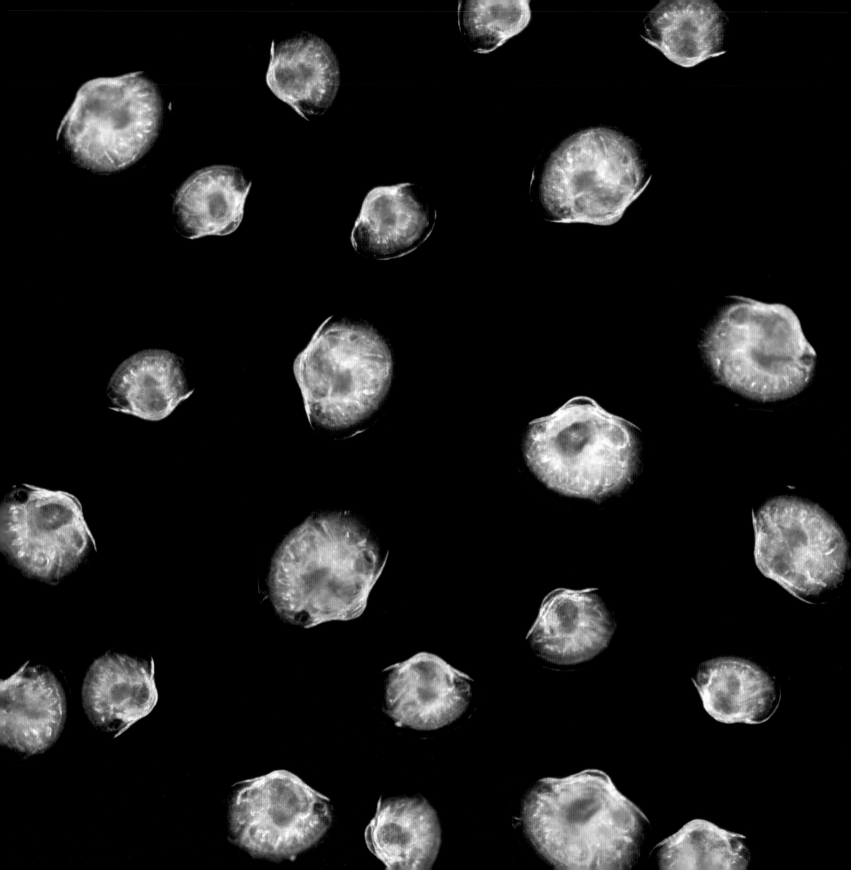

Fish eggs

Most fish lay eggs that float in the plankton. The yolk sac and single oil droplet that will fuel the initial growth of each of these sardine larvae is clearly visible in these eggs. When they hatch, the young fish larvae will feed on the plankton, as will the adult sardine. An important part of fisheries management involves counting the number of fish eggs in the plankton each year. These annual egg counts help fishery scientists estimate the size of the spawning stocks of different species.

The eggs of the sardine *Sardina pilchardus*
MAGNIFICATION X 55

A sardine egg
MAGNIFICATION X 100

Young fish

The plankton underpins fish recruitment (the number of fish larvae that survive to become juveniles and adults), and so the plankton influences fisheries everywhere. After gaining initial nourishment from their egg sac immediately after hatching, juvenile fish like this young dragonet feed on plankton such as small copepods. Consequently, the seasonal timing, distribution, and abundance of the plankton is critical to the growth of larval fish. For example, in the North Pacific, scientists have shown that copepods found in more southern waters occur further north in warm years and the peak biomass of the two important copepods *Neocalanus plumchrus* and *Neocalanus flamingeri* occurs earlier and for a shorter duration when the sea is warmer. Therefore, not only does the quantity of copepods change with sea temperature, but also their composition and perhaps nutritional quality, with possible effects not only on fisheries but also on the migration and breeding patterns of the many seabirds that also depend on the plankton.

A dragonet larva, *Callionymus lyra*
MAGNIFICATION X 110

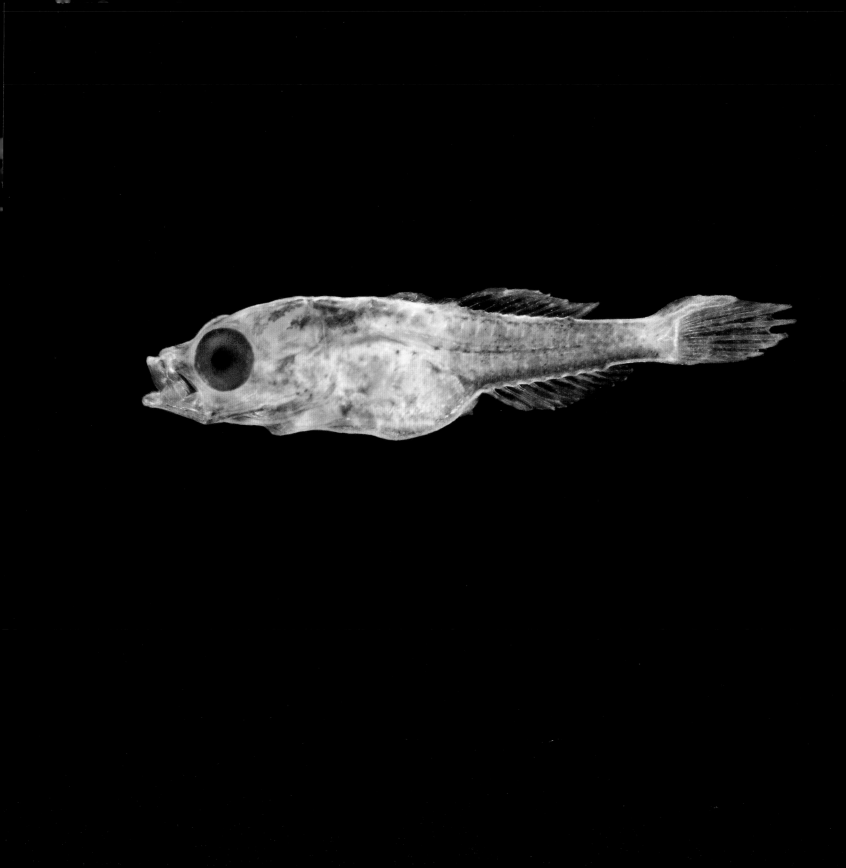

Remarkable metamorphosis

This planktonic larva of the sand sole, *Pegusa lascaris*, that lives its adult life on the sea bed, shows a remarkable metamorphosis. The adult sand sole swims on its side and so it is a type of fish commonly called a flatfish. However, like other fish larvae, the sand sole larva is initially round in shape with one eye on either side of its head. As a flatfish larva develops, one of the eyes migrates slowly over the top of the head to the other side so that they are both on the same side of its body when the juvenile leaves the plankton for its life on the sea floor. In the *P. lascaris*, the right eye migrates to the left side of the head. The gut of the larval flatfish also twists through 50 degrees and the side of the animal that will face upwards on the sea bed develops pigmentation to provide camouflage against the sea floor.

The larva of the sand sole *Pegusa lascaris*

MAGNIFICATION X 80

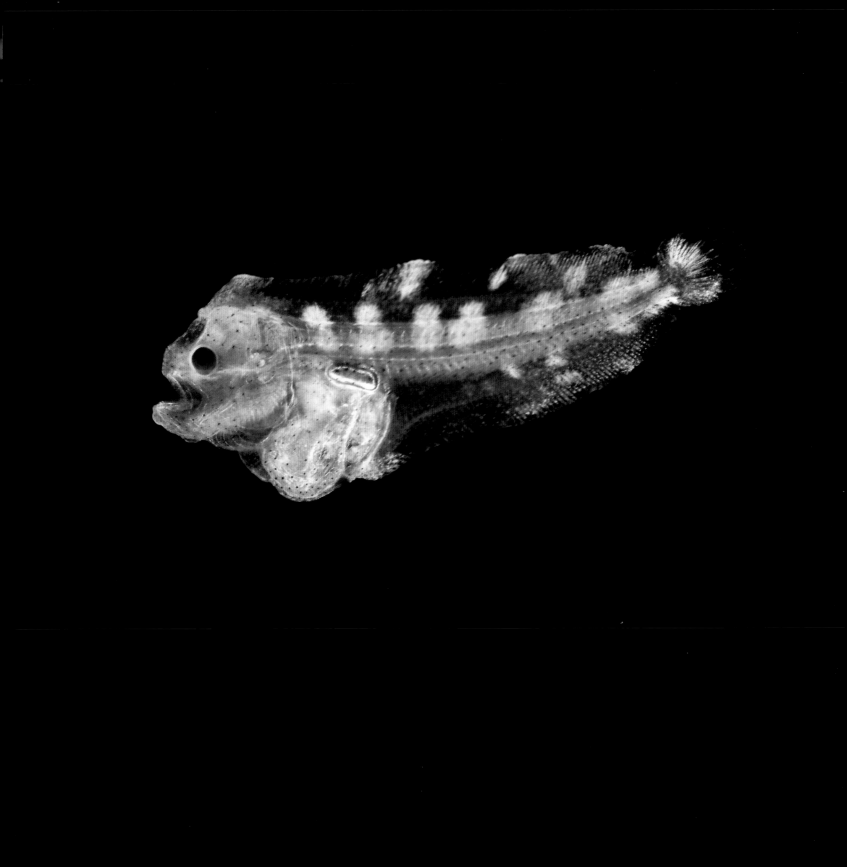

Eyeing up the plankton

The eye of this young rockling fish will help it find its food and avoid predators during its short life in the plankton, before it swims to the sea bed where it will live its adult life beneath a rock. The supply of plankton food and predation by other zooplankton – such as amphipods, chaetognaths, siphonophores and jellyfish – are critical factors influencing larval fish survival and they interact through the phenomenon known as size dependent mortality. The amount and quality of the phytoplankton and zooplankton food available to fish larvae will determine how fast they grow, which in turn determines how long they spend in the plankton where they are vulnerable to their zooplankton predators.

The eye of a juvenile rockling

MAGNIFICATION X 220

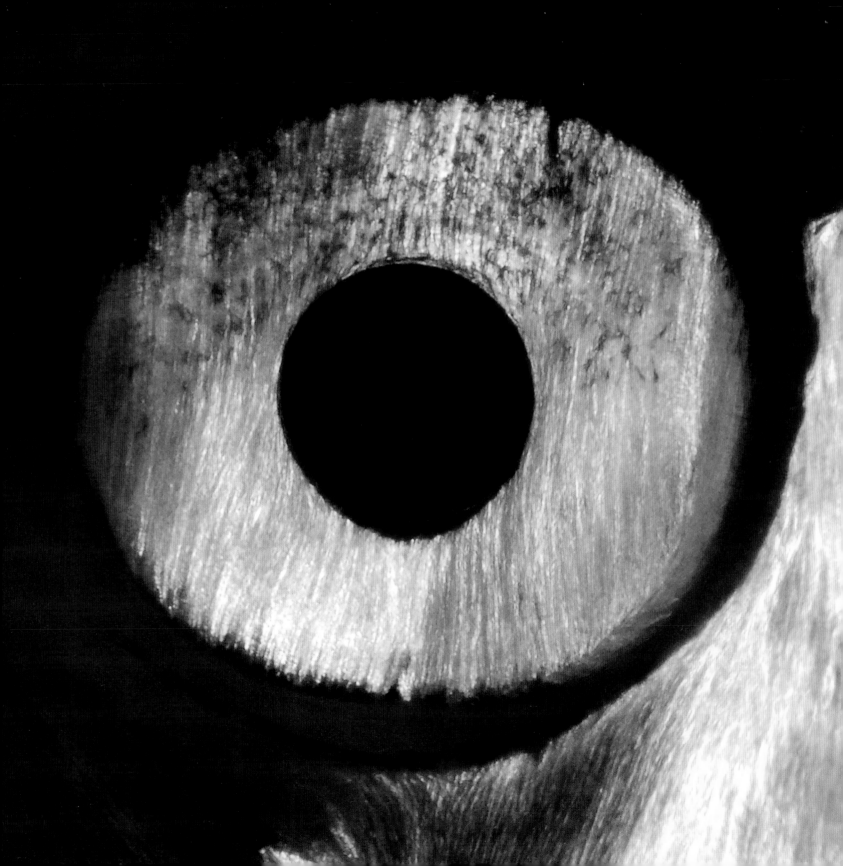

Death and decay

Death and decay completes the cycle of life in the plankton. As the dead plankton, their liquid wastes and faecal pellets are decomposed by marine bacteria, their remains sink slowly to the sea floor. This steady 'rain' of organic material – sometimes called 'marine snow' – removes carbon from the surface waters to the deep ocean in a process that is known as the 'biological carbon pump'. The carbon in the plankton is derived originally from the photosynthetic conversion by phytoplankton of carbon dioxide that is dissolved in the sea from the atmosphere and from *in situ* respiration. Since the phytoplankton account for about 50% of global photosynthesis, and carbon dioxide is a greenhouse gas, the plankton have an important influence on climate. Some of the organic material that reaches the sea floor becomes incorporated into sediments and over geological time scales these deposits created the Earth's oil and gas reserves. Our burning of fossil fuels is returning this carbon to the atmosphere far faster than it is being sequestered by the plankton. As a result, the levels of carbon dioxide in the atmosphere have risen over the last 200 years from approximately 280 ppm to about 380 ppm in 2008. This rate of change is 100 times faster than the most recent natural change in this greenhouse gas, which was at the end of the last ice age, and it has resulted in atmospheric carbon dioxide levels that are now 27% higher than at any point in the last 800,000 years.

Decaying faecal pellets and plankton

MAGNIFICATION X 65

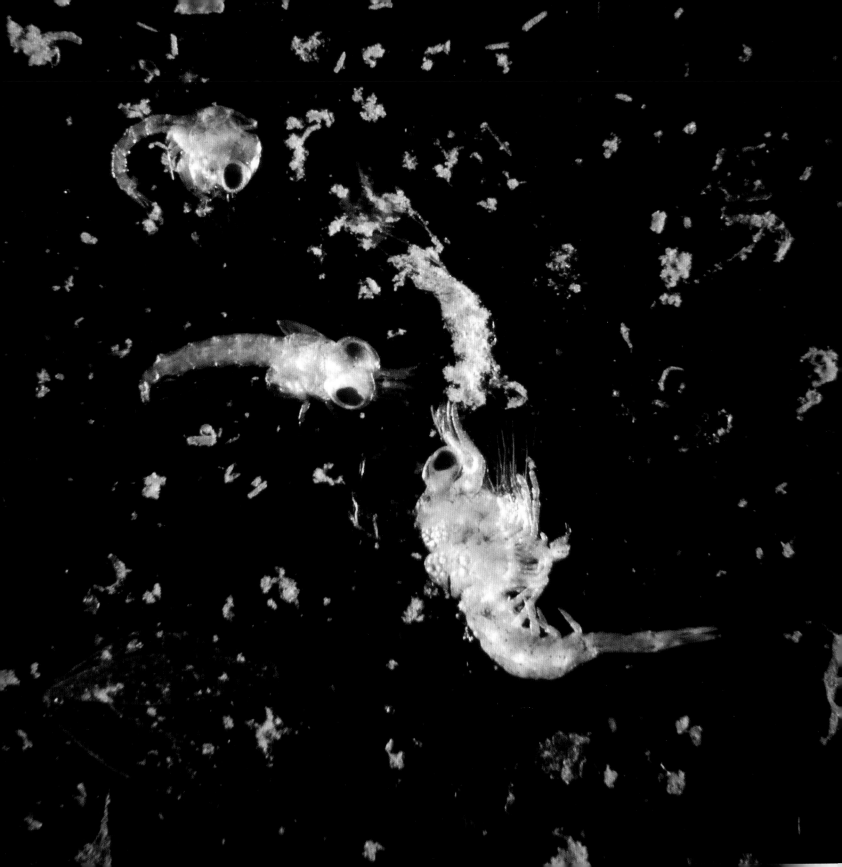

Changing sea temperature

Living in the surface of the sea the plankton are sensitive to changes in sea surface temperature. As a consequence of the anthropogenic (manmade) increase in the greenhouse gas carbon dioxide in the atmosphere, climate models predict that the global average temperature will rise by 0.4° C over the next 20 years. Global sea surface temperatures are already 1°C higher than 150 years ago and coincident changes are being observed in the abundance, distributions and seasonal timing of the plankton in both the Atlantic and Pacific Oceans. In the Northeast Atlantic, warm-water copepods have advanced northwards by approximately 1,200 km over the period 1960–2005 and cold-water species have retreated towards the North Pole. One copepod that has changed its distribution, moving northward in the Northeast Atlantic, is the temperate copepod *Corycaeus anglicus*.

Females of the temperate copepod *Corycaeus anglicus* **with egg masses**
MAGNIFICATION X 100

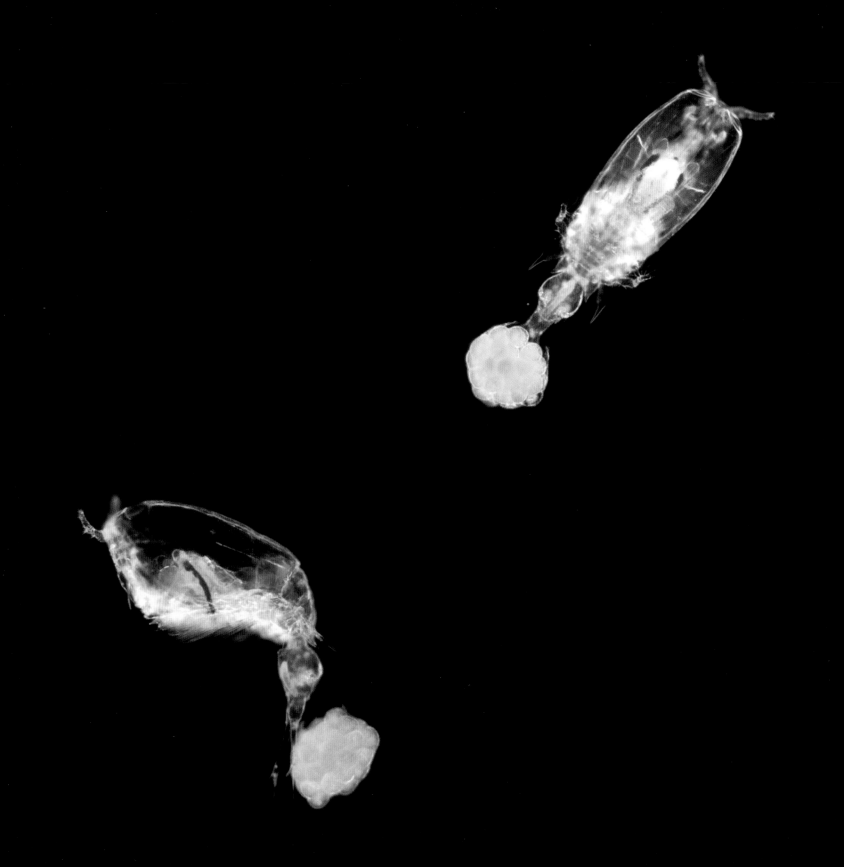

Summing up

The plankton provides one of our planet's most important life-support systems. The photosynthetic phytoplankton, by converting carbon dioxide into organic carbon compounds, underpin the whole marine food chain. It is this plankton food web – from phytoplankton to zooplankton, to fish and whales, the largest mammals ever to live on Earth, and from the seabirds in the sky to the crabs, worms and starfish of the sea bed – that helps to create the spectacular diversity of marine life. Even the deepest of hydrothermal-vent ecosystems, once considered to be independent of the plankton food web, are now thought to be influenced by plankton productivity thousands of metres above. By the continuous drawdown of carbon from the sea surface to the deep ocean as the dead bodies of plankton and other marine organisms sink to the sea floor, the plankton play a central in the global carbon cycle. Despite the global importance of the plankton, and even though they live at the surface of the sea, these remarkable creatures are still very much a secret world. Many of the incredible adaptations for life revealed in the pictures of this book are poorly understood and the extent of their true biodiversity remains unknown. As the plankton habitat alters as sea surface temperatures warm due to current global climate change, the plankton are shifting their distributions with ramifications for the whole marine food web and perhaps the ecology of our planet.

The larvae of a sea anemone

MAGNIFICATION X 100

Studying the plankton

One of the first and simplest devices used to study the plankton in its environment was the Secchi disk, named after the scientist, Pietro Angelo Secchi, who invented this tool in 1865. The latest version of the Secchi disk comprises a white plastic disk of 15 cm radius that is attached to a length of rope marked in 50 cm intervals. When the weighted disk is lowered into the water, the depth at which it disappears from sight is called the Secchi depth. The Secchi depth provides a measure of the turbidity (cloudiness) of the water; the clearer the water the greater the depth to which light penetrates and so the greater is the Secchi depth. Phytoplankton affect water turbidity so the Secchi depth can be used to monitor the development and progression of the spring phytoplankton bloom. Further, as the Secchi depth gives an indication of the depth to which light can penetrate, it estimates the depth to which phytoplankton can grow in the water column.

Plankton nets

To collect samples of plankton for study in the laboratory scientists typically use conical-shaped plankton nets. The first plankton nets are believed to date to the early 19th century and were developed to sample crab and barnacle larvae by a scientist called J.V. Thomson. Today, deployed from research vessels, the nets may be hauled vertically through the water to obtain a depth profile of the plankton, or they may be towed obliquely just beneath the surface. As the net is towed the plankton are strained from the water and washed down into a collection bottle at the end. One of the first accounts of plankton sampling by net was given by Charles Darwin who collected plankton on the *Beagle* cruise in 1832 during a voyage between Teneriffe and the Cape Verde islands. On 10 January 1832, Darwin wrote

SAMPLING PLANKTON
Since the beginning of modern-day marine biology in the 19th century, scientists have been interested in the tiny creatures living beneath the surface of the sea. This photograph from the early 20th century shows two technicians from the UK's Marine Biological Association retrieving a plankton sample from the same waters off Plymouth where plankton samples were collected for this book. The device they are using is a simple plankton ring net with a rope bridle and a collection pot at the cod-end. The plankton will be poured into the jars seen in the wicker baskets.

A MODERN PLANKTON NET
Retrieving a plankton sample from a WP-2 plankton net that has been hauled vertically through the water column. The WP-2 net is a standard net for plankton collection; it has a mouth of approximately 60 cm diameter and a mesh size of 0.2 mm. These two sea-going technicians onboard the research vessel, *MBA Sepia*, have retrieved a plankton sample collected in the coastal waters of Plymouth, UK. This WP-2 net, along with a bongo net system and a 1 m fish-larva net were used to collect the plankton photographed in this book.

in his *Beagle Diary*, "I proved to day the utility
of a contrivance which will afford me many
hours of amusement and work. — it is a bag
four feet deep, made of bunting, and attached
to semicircular bow this by lines is kept upright,
and dragged behind the vessel — this evening it
brought up a mass of small animals, and tomorrow
I look forward to a greater harvest". Today,
plankton nets can range in mesh size from
anything between 10 μm (0.01 mm) to 1.0 mm,
and they may be made from silk or nylon
monofilament mesh fabric. The two most common
nets employed in combination for phytoplankton
and zooplankton sampling are those with fine,
approximately 20 μm (0.02 mm), and coarse,
approximately 150 μm (0.15 mm), mesh sizes.
While the simplest nets comprise a tapering tube
of gauze leading to a collecting bottle, more
sophisticated nets also exist. Such nets may
involve a reverse tapering entrance, which reduces
the amount of water entering the net and so
improves filtering, or they may have a mechanism
to close the opening so that plankton are collected
from a particular depth.

Hi-tech survey systems
While simple plankton nets and Secchi disks
have served marine science for over a hundred
years, many other technologies have also been
developed to help study the plankton and their
environment. The Undulating Oceanographic
Recorder (UOR) is a sampling device that is
towed behind research ships and that can be
equipped with various sensors to measure the
biological and physical properties of the sea
surface. The UOR has a variable pitch hydrofoil
that causes it to rise and fall in the surface
waters between depths of 15 and 70 m over a
variable undulation distance of between 3 and
30 km. Depending on the sensors that are fitted

on the UOR it can record parameters such as
salinity, temperature, the amount of down–
welling and up–welling light, and chlorophyll
levels in the seawater. The Longhurst–Hardy
Plankton Recorder (LHPR) named after the two
scientists who invented it, is a towed
zooplankton sampling net mounted in a frame.
The LHPR has a battery powered mechanical
unit at the cod-end within which is a long strip
of mesh that advances in a stepped manner
every 30 seconds. In this way, the LHPR can
record a series of discrete zooplankton samples.

ALISTER HARDY
Sir Alister Hardy FRS, the inventor of the Continuous
Plankton Recorder, is pictured here holding the
machine's internal plankton sampling mechanism.
Hardy invented his first plankton recorder to collect
plankton during the Discovery Expedition to
Antarctica from 1925 to 1927. Today, almost
unchanged in design, the machines are towed
behind ships throughout the world's oceans. The
plankton samples that these machines collect are
helping scientists understand the plankton and the
intricacies of the marine food web.

Like the UOR, the LHPR can be equipped with various sensors to provide a description of the physical habitat of the plankton at the precise location where they are caught. In addition to towed samplers, static sediment traps can also be deployed vertically at different depths in the water column where they collect decaying planktonic matter to help determine the flux of nutrients to the sea bed.

Sonar and video are also used to visualise the plankton *in situ*. Ship-borne, acoustic Doppler current profilers (ADCPs) emit sound pulses that are scattered by the plankton in the water column to reveal their horizontal and vertical distribution. The ADCP can detect both individual zooplankton and aggregations, and so successive measurements taken over time help map zooplankton vertical migration. High-definition video camera systems can photograph individual plankton in their natural environment. One such system is the PICASSO (Plankton Investigatory Collaborating Autonomous Survey System Operon) underwater vehicle developed by the Japan Agency for Marine–Earth Science and Technology (JAMSTEC). These autonomous underwater camera systems are revealing a whole new world of gelatinous planktonic organisms that are so fragile they are often destroyed by conventional sampling with plankton nets. By observing the plankton *in situ* with cameras, a valuable insight is also being gained into how several planktonic organisms often live in association with each other; these natural associations are usually disrupted by conventional sampling methods.

Today, plankton communities are also being monitored from space. Phytoplankton, because they contain photosynthetic and other pigments, colour the water when they bloom. This bio-optical property of the sea surface is measured

THE CPR SURVEY

A fleet of CPRs wait to go to sea from the workshop of the Sir Alister Hardy Foundation for Ocean Science. The modern-day CPR is towed at a constant depth of 7 m behind a merchant ship and the plankton are retained on a moving band of silk gauze. Today, several CPRs are at sea at any one time as part of the CPR survey that Alister Hardy founded in the 1930s. By 2008, CPRs had been towed over 5 million nautical miles and over 190,000 plankton samples had been analysed. The CPR database is the largest marine biological time-series in the world.

TESTING THE WATER

A conductivity, temperature and depth (CTD) array and rosette of Niskin bottles is used to measure salinity and sea temperature, and to collect water samples for laboratory analysis. Scientists need to appreciate the physical and chemical characteristics of the water to understand the plankton habitat. In the laboratory the water is analysed for nutrients, oxygen and other gases, which along with the CTD data, provide a physical and chemical vertical profile of the water column that is often complemented today by satellite images of the surface.

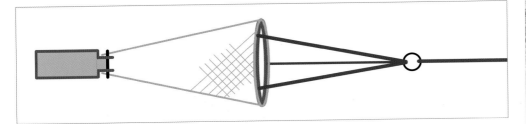

BASIC PLANKTON NET
You can make a simple plankton net from household items: a wire ring, which can be made from a coat hanger; about 9 m of strong cord; a stocking with the toe removed; a 500 ml clear plastic drink bottle; and a cable tie.

TWIN NETS
Twin plankton ring nets returning to the ship after being towed obliquely from a depth of 40 m to the surface over a period of about 20 minutes. Several of these paired nets can be connected at intervals to a weighted line so they can collect many samples.

by orbiting satellites such as the Sea-viewing Wide Field-of-view Sensor (SeaWiFS) satellite operated by the National Aeronautics and Space Administration (NASA). Provided the sky is cloud free, these orbiting sensors produce a global view of the ever-changing phytoplankton at a scale of every square kilometre. Together with *in situ* measurements of salinity, temperature and nutrient profiles, a knowledge of fronts and ocean currents and plankton sampling by ships, these satellite images help build up an understanding of the temporal variability of the plankton over large geographic areas.

Plankton DNA

The most recent development in plankton ecology is the application of modern molecular genetic methods involving the study of the DNA of planktonic organisms. The DNA of plankton can be used to help identify them, to help understand their evolution and to help determine whether species vary genetically from place to place. Currently, an ambitious project is the

Census of Marine Life (CoML), which aims to construct a database containing a DNA sequence from every marine species. It is hoped that this international project will help resolve some of the mysteries of the plankton; for example, how many species there are or what adult creature some of the still unidentified larvae, such as the facetotectan larva, eventually become. DNA sequences have already proved useful in helping to understand the biodiversity of the bacterioplankton and resolve what was known as 'the great plate count anomaly'. When seawater is studied at high magnification under the microscope it appears to be teeming with free-living bacteria at densities of up to 10 million cells per ml. However, when a sample of seawater was cultured in the laboratory using traditional methods, only a few colonies would grow in a petri dish: hence, the great plate count anomaly. By filtering seawater to collect the bacterioplankton and then extracting their DNA directly, molecular methods have uncovered a vast biodiversity of bacterioplankton (up to

10,000 different types per ml) creating a whole new field of plankton research.

The Continuous Plankton Recorder

Arguably however, perhaps just one scientist, Alister Hardy, a fisheries biologist in the UK in the 1920s, has contributed more than anyone to our scientific understanding of the plankton and their environment. Hardy was interested in why the recruitment of the North Sea Herring, *Clupea harengus*, varied so much from year to year. Hardy's scientific colleagues persuaded him that, if he were ever to understand this conundrum, he would need to understand the herring's food, the plankton, and the intricacies of the planktonic food web.

In the 1920s, there was no easy way to study the plankton regularly over an area as large as the North Sea. Hardy therefore invented a machine that he called the Continuous Plankton Recorder (CPR). The CPR is a plankton sampling instrument that is towed at a constant depth of 7 m behind a ship. Seawater enters the front of the machine through a small aperture and inside the plankton are retained on a moving band of silk gauze that is slowly wound into a formalin storage tank where the plankton are preserved. Six metres of silk equals a tow of 500 nautical miles and, as the silk travels through the mechanism continuously, it provides a record of the plankton along the whole route of the ship. When the tow-route finishes, the CPR is returned to the laboratory where the silk is removed and divided into sections each representing 10 nautical miles. The plankton samples are then identified and counted.

Although Hardy invented the CPR to help understand herring recruitment, the prototype CPR was first deployed in the Southern Ocean to sample Antarctic krill during the "Discovery" cruises of 1925-27. After these 'sea trials', Hardy modified the design slightly and started collecting plankton in the North Sea. Here, the CPR proved so successful that Hardy established the CPR survey in 1932. Except for a break during World War II, the survey has operated in the North Sea and North Atlantic ever since with the first transatlantic route towed between Reykjavik and Newfoundland in 1959. Due to his contributions to marine science, Hardy became Sir Alister Hardy FRS. Today, the CPR survey is operated by the Sir Alister Hardy Foundation for Ocean Science in Plymouth, UK, and each month, many merchant ships voluntarily tow CPRs on their normal trading routes. Since the start of the CPR survey, more than 5 million nautical miles of the North Sea and North Atlantic, and more recently the Southern and Pacific Oceans have been sampled. Scientists across the world now use the results of this ongoing, remarkable survey to help understand the plankton and the changes in the environment that their changing distributions and abundance may signal.

ALGAL CULTURES

In laboratories around the world, scientists use culture collections to help their research on phytoplankton. While these cultures are quite small volumes, microalgae can be grown on a vast scale and their culture is now being investigated as a source of biofuels, and as a method of carbon capture at fossil fuel power stations.

WITHOUT THE PLANKTON THERE WOULD BE NO FISH IN THE SEA

Index

Further reading

Books

Berner, R. A. (2004) *The Phanerozoic carbon cycle*. Oxford University Press, Oxford.

Bougis, P. (1976) *Marine plankton ecology*. North-Holland Publishing Company/American Elsevier Publishing Company, Amsterdam and New York.

Darwin, C. R. (1851). Living Cirripedia, A monograph on the sub-class Cirripedia, with figures of all the species. The Lepadidae; or, pedunculated cirripedes. Volume 1, The Ray Society, London.

Darwin, C. R. (1854) Living Cirripedia, The Balanidae, (or sessile cirripedes); the Verrucidae. Volume 2, The Ray Society, London.

Garstang, W. (1951) *Larval forms and other zoological verses*. Blackwell, Oxford.

Grahame, J. (1987) *Plankton and fisheries*. Edward Arnold, London.

Granéli, E. and Turner, J. T. (eds.) (2006) *Ecology of harmful algae*. *Ecological Studies: Analysis and Synthesis* 189. Springer Verlag, Berlin.

Hardy, A. C. (1956) *The open sea: its natural history: the world of plankton*. Collins, London.

Hardy, A. C. (1959). *The open sea: its natural history. Pt. II. fish and fisheries, with chapters on whales, turtles and animals of the sea floor*. Collins, London.

Harris, R., Wiebe, P., Lenz, J., Skjoldal, H-R. and Huntley, M. (Eds.) (2000) *ICES zooplankton methodology manual*. Academic Press, San Diego, California.

Hooke, R. (1665) *Micrographia: or some physiological descriptions of minute bodies made by magnifying glasses*. John Martyn and James Allestry, London.

IPCC (2007) *Climate change 2007: the physical science basis*. Cambridge University Press, Cambridge.

Johnson, W. S. and Allen, D. M. (2005) *Zooplankton of the Atlantic and Gulf coasts: a guide to their identification and ecology*. The Johns Hopkins University Press, Baltimore.

Keynes, R. D. ed. (2001) Charles Darwin's Beagle Diary. Cambridge University Press, Cambridge.

Kiørboe, T. (2008) *A mechanistic approach to plankton ecology*. Princeton University Press, Princeton, New Jersey.

Laybourn-Parry, J. (1992) *Protozoan plankton ecology*. Chapman and Hall, London and New York.

McEdward, L. (ed.) (1995) *Ecology of marine invertebrate larvae*. CRC Press, Boca Raton, Florida.

Newell, G. E. and Newell, R. C. (1977) *Marine plankton: a practical guide*. Hutchinson, London.

Prothero, D. R. (2004) *Bringing fossils to life: an introduction to paleobiology*. McGraw Hill, Boston, Massachussetts.

Railkin A. I. (2004) Marine Biofouling: colonization processes and defenses. CRC Press, Washington, DC.

Royal Society (2005) *Ocean acidification due to increasing atmospheric carbon dioxide*. The Royal Society, London.

Shimomura, O. (2006) *Bioluminescence: Chemical Principles And Methods*. World Scientific Publishing Company, Singapore.

Smith, D. L. and Johnson, K. B. (1996). *A guide to marine coastal plankton and marine invertebate larvae*. Kendal/Hunt Publishing Company, Dubuque, Iowa.

Suthers, I. M. and Rissik, D. (eds.) (2009) *Plankton: a guide to their ecology and monitoring for water quality*. CSIRO Publishing, Collingwood, Australia.

Todd, C. D., Laverack, M. S. and Boxshall, G. A. (1996) *Coastal marine zooplankton: a practical manual for students*. Cambridge University Press, Cambridge.

Tomas, C. R. (ed.) (1997) *Identifying marine phytoplankton*. Academic Press, San Diego, California.

Vezzulli, L., Dowland, P. S., Reid, P. C. and Hylton, E. K. (2007) Gridded database browser of North Sea plankton, version 1.1: Fifty four years (1948–2001) of monthly plankton abundance from the Continuous Plankton Recorder (CPR) survey. Sir Alister Hardy Foundation for Ocean Science. Available from http://cpr.cisnr.org.

Wickstead, J. H. (1976) *Marine zooplankton*. Edward Arnold, London.

Wimpenny, R. S. (1966) *The plankton of the sea*. Faber and Faber, London.

Scientific literature

Abrahams, M. V. and Townsend, L. D. (1993) Bioluminescence in dinoflagellates: a test of the burglar alarm hypothesis. *Ecology* **74**: 258–260.

Atkinson, A., Siegel, V., Pakhomov, E. and Rothery. P. (2004) Long-term decline in krill stock and increase in salps within the Southern Ocean. *Nature* **432**: 100–103.

Barkai, A. and McQuaid, C. (1988) Predator-prey role reversal in a marine benthic ecosystem. *Science* **242**: 62–64.

Batten, S. D. and Welch, D. W. (2004) Changes in oceanic zooplankton populations in the North-east Pacific associated with the possible climatic regime shift of 1998/99. *Deep Sea Research II* **51**: 863–873.

Beaugrand, G. and Kirby, R. R. (2009) Spatial changes in the sensitivity of the Atlantic cod to climate-driven effects in the plankton. *Climate Research* **41**, 15–19.

Beaugrand, G., Reid, P. C., Ibanez, F., Lindley J. A. and Edwards, M. (2002) Reorganization of North Atlantic marine copepod biodiversity and climate. *Science* **296**: 1692–1694.

Beaugrand, G., Brander, K. M., Lindley, J. A., Souissi, S. and Reid, P. C. (2003) Plankton effect on cod recruitment in the North Sea. *Nature* **426**: 661–664.

Blades, P. I. (1977) Mating behavior of *Centropages typicus* (Copepoda: Calanoida). *Marine Biology* **40**: 57–64.

Blankenship, R. E. (1992) Origin and early evolution of photosynthesis. *Photosynthesis Research* **33**: 91–111.

Buskey, E. J. and Hartline, D. K. (2003) High-speed video analysis of the escape responses of the copepod *Acartia tonsa* to shadows. *Biological Bulletin* **204**: 28–37.

Casini M, Lövgren J, Hjelm J, Cardinale M, Molinero J-C. and Kornilovs, G. (2007) Multi-level trophic cascades in a heavily exploited open marine ecosystem. *Proceedings of the Royal Society B* **275**: 1793–1801.

Charlson, R. J., Lovelock, J. E., Andrease, M. O. and Warren, S. G. (1987) Oceanic phytoplankton, atmospheric sulfur, cloud albedo and climate, a geophysical feedback. *Nature* **326**: 655–661.

Conover, R. J. (1968) Zooplankton – Life in a nutritionally dilute environment. *American Zoologist* 8: 107–118.

Daskalov, G. M., Grishin, A. N., Rodionov, S. and Mihneva, V. (2007) Trophic cascades triggered by overfishing reveal possible mechanisms of ecosystem regime shifts. *Proceedings of the National Academy of Sciences USA* 104: 10518–10523.

De Deckker, P. (2004) On the celestite-secreting Acantharia and their effect on seawater strontium to calcium ratios. *Hydrobiologia* **517**: 1–13.

Dreanno, C., Kirby, R. R. and Clare, A. S. (2006) Smelly feet are not always a bad thing: the relationship between cyprid footprint protein and the barnacle settlement pheromone. *Biology Letters* 2: 423–425.

Dreanno, C., Matsumura, K., Dohmae, N., Takio, K., Hirota, H., Kirby, R. R. and Clare, A. S. (2006) An $^a_2$-macroglobulin-like protein is the cue to gregarious settlement of the barnacle, *Balanus amphitrite*. *Proceedings of the National Academy of Sciences USA* **103**: 14396–14401.

Edwards, M. and Richardson, A. J. (2004) Impact of climate change on marine pelagic phenology and trophic mismatch. *Nature* **430**: 881–884.

Frank, K. T., Petrie, B., Choi, J. S. and Leggett W. C. (2005) Trophic cascades in a formerly cod-dominated ecosystem. *Science* **308**: 1621–1623.

Frouin, R. and Iacobellis S. F. (2002) Influence of phytoplankton on the global radiation budget. *Journal of Geophysical Research* **107**: No. D19, 4377, doi:10.1029/2001JD000562.

Goetze, E. (2003) Cryptic speciation on the high seas; global phylogenetics of the copepod family Eucalanidae. *Proceedings of the Royal Society of London B* **270**: 2321–2331.

Heath, M. R. (2005) Changes in the structure and function of the North Sea fish foodweb, 1973–2000, and the impacts of fishing and climate. *ICES Journal of Marine Science* **62:** 847–868.

Hunt, B. P. V., Pakhomov, E. A., Hosie, G. W., Siegel, V., Ward, P. and Bernard, K. (2008) Pteropods in Southern Ocean ecosystems. *Progress in Oceanography* **78**: 193–221.

Iglesias-Rodriguez, M. D. and 12 others (2008) Phytoplankton calcification in a high-CO_2 world. *Science*: **320**, 336–340.

Jackson, J. B. C. (2008) Ecological extinction and evolution in the brave new ocean. *Proceedings of the National Academy of Sciences USA* **105**: 11458–11465.

Jacobson, D. M. and Anderson, D. M. (1986) Thecate heterotrophic dinoflagellates: feeding behaviours and mechanisms. *Journal of Phycology* **22**: 249–258.

Jeong, H. J. and Latz, M. I. (1994) Growth and grazing rates of the heterotrophic dinoflagellates *Protoperidinium* spp. On red tide dinoflagellates. *Marine Ecology Progress Series* **106**: 173–185.

Kirby, R. R. and Beaugrand, G. (2009) Trophic amplification of climate warming. *Proceedings of the Royal Society* B **276**, 4095–4103.

Kirby, R. R., Beaugrand, G. and Lindley, J. A. (2008) Climate-induced effects on the meroplankton and the benthic pelagic ecology of the North Sea. *Limnology and Oceanography* **53**: 1805–1815.

Kirby, R. R., Beaugrand, G. and Lindley, J. A. (2009) Synergistic effects of climate and fishing in a marine ecosystem. *Ecosystems* **12**: 548–561.

Liss, P. S., Hatton, A. D., Malin, G., Nightingale, P. D. and Turner, S. M. (1997) Marine sulphur emissions. *Philosophical Transactions of the Royal Society of London B.* **352**: 159–68.

Lüthi, D. and 10 others (2008) High-resolution carbon dioxide concentration record 650,000–800,000 years before present. *Nature* **453**: 379–382.

Lynam, C. P., Gibbons, J., Axelsen, B., Sparks, C. A. J., Coetzee, J., Heywood, B. G. and Brierley, A. S. (2006) Jellyfish overtake fish in a heavily fished ecosystem. *Current Biology* **16**: 492–493.

Mackas, D. L., Batten, S. and Trudel, M. (2007) Effects on zooplankton of a warmer ocean: recent evidence from the Northeast Pacific. *Progress in Oceanography* **75**: 223–252.

Martin, J. H. and 43 others (2002) Testing the iron hypothesis in ecosystems of the equatorial Pacific Ocean. *Nature* **371**: 123 – 129.

Mauchline, J. (1998) The biology of calanoid copepods. *Advances in Marine Biology* **33**.

Miller, S. D., Haddock, S. H. D., Elvidge, C. D. and Lee, T. F. (2005) Detection of a bioluminescent milky sea from space. *Proceedings of the National Academy of Sciences USA* **102**: 14181–14184.

Nishikawa, T., Tarutani, K. and Yamamoto, T. (2008) Nitrate and phosphate uptake kinetics of the harmful diatom *Eucampia zodiacus* Ehrenberg, a causative organism in the bleaching of aquacultured *Porphyra* thalli. *Harmful Algae* **8**: 513–517.

Olson, J. M. (2006) Photosynthesis in the Archean era. *Photosynthesis Research* **88**: 109–117.

Pakhomov, E. A. and Perissinotto, R. (1996) Trophodynamics of the hyperiid amphipod *Themisto gaudichaudi* in the South Georgia region during late austral summer. *Marine Ecology Progress Series* **134**: 91–100.

Pauly, D., Christensen, V., Dalsgaard, J., Froese, R. and Torres, F. Jr. (1998) Fishing down marine food webs. *Science* **279**: 860–863.

Pérez-Losada, M., Høeg, J. T. and Crandall, K. A. (2009) Remarkable convergent evolution in specialized parasitic Thecostraca (Crustacea). *BMC Biology* **7**:15 doi:10.1186/1741-7007-7-15.

Piraino, S., Boero, F., Aeschbach, B. and Schmid, V. (1996) Reversing the life cycle: medusae transforming into polyps and cell transdifferentiation in *Turritopsis nutricula* (Cnidaria, Hydrozoa). *The Biological Bulletin* **190**: 302–312.

Pollard, R. T. and 33 others. (2009) Southern Ocean deep-water carbon export enhanced by natural iron fertilization. *Nature* **457**: 577–580

Purcell, J. E. (2005) Climate effects on formation of jellyfish and ctenophore blooms: a review. *Journal of the Marine Biological Association of the UK* **85**: 461–476.

Reid, P. C., Johns, D. G., Edwards, M., Starr, M., Poulin, M. and Snoeijs, P. (2007) A biological consequence of reducing Arctic ice cover: arrival of the Pacific diatom *Neodenticula seminae* in the North Atlantic for the first time in 800 000 years. *Global Change Biology* **13**: 1910–1921.

Rouse, G. W. (1999) Trochophore concepts: Ciliary bands and the evolution of larvae in spiralian Metazoa. *Biological Journal of the Linnean Society* **66**: 411–464.

Ruiz, G. M., Carlton, J. T., Grosholz, E. D. and Heines, A. H. (1997) Global invasions of marine and estuarine habitats by non-indigenous species: mechanisms, extent and consequences. *American Zoologist* **37**: 621–632.

Scheffer, M., Carpenter, S. and deYoung, B. (2005) Cascading effects of overfishing marine systems. *Trends in Ecology and Evolution* **20**: 570–581.

Shepherd, J. G. and Cushing, D. H. (1980) A mechanism for density-dependent survival of larval fish as a basis of a stock-recruitment relationship. *Conseil International pour l'Exploration de la Mer* **39**: 160–167.

Wanless, S., Harris, M. P., Redman, P. and Speakman, J. R. (2005) Low energy values of fish as a probable cause of a major seabird breeding failure in the North Sea. *Marine Ecology Progress Series* **294**: 1–8.

Watson, A. J. (1997) Volcanic iron, CO2, ocean productivity and climate. *Nature* **385**: 587–588.

Acknowledgments

The author would like to thank the crews of the research ships MBA Sepia and Plymouth Quest who collected plankton from the sea off Plymouth, UK and D Conway, M Kendall, J Lindley and R Pipe for their help with plankton identification.

Studio Cactus would like to thank Kate Byrne for proof reading, Penelope Kent for the index, and the following photographers for permission to reproduce copyright material: idreamphoto 8 (bl); Gail Johnson 8 (br); Witold Kaszkin 15; Atlaspix 18.